ELECTRICIAN'S
Technical Reference

Motor Controls

ELECTRICIAN'S
Technical Reference

Motor Controls

David R. Carpenter

Africa • Australia • Canada • Denmark • Japan • Mexico • New Zealand • Philippines
Puerto Rico • Singapore • Spain • United Kingdom • United States

Notice to the Reader

Publisher does not warrant or guarantee any of the products described herein or perform any independent analysis in connection with any of the product information contained herein. Publisher does not assume, and expressly disclaims, any obligation to obtain and include information other than that provided to it by the manufacturer.

The reader is expressly warned to consider and adopt all safety precautions that might be indicated by the activities herein and to avoid all potential hazards. By following the instructions contained herein, the reader willingly assumes all risks in connection with such instructions.

The publisher makes no representation or warranties of any kind, including but not limited to, the warranties of fitness for particular purpose or merchantability, nor are any such representations implied with respect to the material set forth herein, and the publisher takes no responsibility with respect to such material. The publisher shall not be liable for any special, consequential, or exemplary damages resulting, in whole or part, from the readers' use of, or reliance upon, this material.

Delmar Staff

Business Unit Director: Alar Elken
Executive Editor: Sandy Clark
Acquisitions Editor: Mark Huth
Developmental Editor: Jeanne Mesick
Editorial Assistant: Dawn Daugherty
Executive Marketing Manager: Maura Theriault
Channel Manager: Mona Caron
Marketing Coordinator: Kasey Young
Executive Production Manager: Mary Ellen Black
Project Editor: Barbara L. Diaz
Art Director: Nicole Reamer

COPYRIGHT © 2000
Delmar is a division of Thomson Learning. The Thomson Learning logo is a registered trademark used herein under license.

Printed in the United States of America

1 2 3 4 5 6 7 8 9 10 XXX 05 04 03 02 01 00

For more information, contact Delmar at
3 Columbia Circle, PO Box 15015, Albany, New York 12212-5015; or find us on the World Wide Web at *http://www.delmar.com*

Asia

Thomson Learning
60 Albert Street, #15-01
Albert Complex
Singapore 189969

Australia/New Zealand

Nelson/Thomson Learning
102 Dodds Street
South Melbourne, Victoria
3205
Australia

Canada

Nelson/Thomson Learning
1120 Birchmont Road
Scarborough, Ontario
Canada M1K 5G4

International Headquarters

Thomson Learning
International Division
290 Harbor Drive, 2nd Floor
Stamford, CT 06902-7477

Japan

Thomson Learning
Palaceside Building 5F
1-1-1 Hitotsubashi, Chiyoda-ku
Tokyo 100 0003
Japan

Latin America

Thomson Learning
Seneca, 53
Colonia Polanco
11560 Mexico D.F. Mexico

Spain

Thomson Learning
Calle Magallanes, 25
28015-Madrid
Espana

UK/Europe/Middle East

Thomson Learning
Berkshire House
168-173 High Holborn
London
WC1V 7AA United Kingdom

Thomas Nelson & Sons Ltd.
Nelson House
Mayfield Road
Walton-on-Thames
KT 12 5PL United Kingdom

Contents

Preface

Everyone responsible for industrial and commercial facilities knows the reliability of any facility depends on its motors. It is almost impossible to imagine any facility, process, or entity that does not involve electric motors somewhere or somehow. Therefore, it is obvious that proper design, understanding, and maintenance of motor controls is essential to the dependability of an operation.

When essential controls fail to function properly, lost production and lost revenues may run into hundreds of thousands of dollars. Failure of a simple conveyor can hold up an entire plant and idle thousands of workers. Downtime is costly in a variety of processes—pharmaceutical, steel, heavy machinery, semiconductor manufacturing, and laboratory research work. Machines that incorporate several motors and operate at less then 100 percent capacity show a percentage of lost revenue that can amount to a large financial loss when the malfunction occurs for long periods of time.

Another cost occurs when those performing the maintenance, design and installation of motor controls do not understand the proper operation and application of motor control systems. This problem may go unnoticed for some time but will result in a large net loss.

This book is written with the electrician, instrument mechanic, and engineer in mind. It should prove to be a valuable resource for those involved with motor application, troubleshooting and training. The text will serve to explain the charts, tables and illustrations that are to be used as quick references for field applications.

The text is designed to build on one concept at a time. Some experienced readers may skip the first four chapters and go to the troubleshooting guides, component functions, and techniques. Chapter 1 covers the general principles of motor control. Chapter 2 illustrates the different types of control methods. Chapter 3 gives information on magnetic clutches and drives. Chapter 4 covers digital logic. Chapter 5 discusses starter design and troubleshooting. Chapter 6 provides information on control relays and how they are used in field applications. Chapter 7 outlines switches and sensor theory application. Chapter 8 describes control transformers. Chapter 9 illustrates the operation and application of solenoid valves. Chapter 10 aids in understanding how to design, read, and troubleshoot wiring diagrams and schematics. Chapter 11 covers the operation and design of DC motor controls. Chapter 12 covers the operation and design of AC motor controls. Chapter 13 illustrates the theory, design, and troubleshooting of drives. Chapter 14 discusses motor control center installation, operation, and maintenance. Chapter 15 outlines motor and motor control protection.

The appendices, illustrations, and tables are quick-reference guides that should prove useful for those working with motors and controls. This is an ideal book for those who need a quick reference to the operation of motor control components.

Acknowledgments

The author and Delmar Publishers wish to acknowledge and thank the members of our editorial panel for their help in developing the concept of this series and reviewing the manuscript.

A. J. Pearson, Director
National Joint Apprenticeship and Training Committee
Upper Marlboro, MD

Robert Baird
Independent Electrical Contractors Association
Alexandria, VA

Brooke Stauffer
Bethesda, MD

P. C. Paul Howard
LaPorte, TX

Jeffrey Lew
Purdue University
West Lafayette, IN

Stephen Falavolito
Dean Institute of Technology
Pittsburgh, PA

Mike Flowers
Tennesee Technology Center
Paris, TN

Michael Freiner
Florida Electrical Apprenticeship Training
Orlando, FL

We would also like to acknowledge and thank Siemens Energy and Automation, Inc. and Square D Company for allowing us generous use of many illustrations and material.

CHAPTER

1

Introduction to Motor Control

The purpose of this chapter is to define terms used for motor control and its application. Understanding these terms is essential to the troubleshooting, designing, and application of motor control. Motor control in its simplest form is stopping, starting or speed control of a motor for a given operation. Devices such as switches, sensors, starters, contactors and relays accomplish this control and are discussed in greater detail later in this text.

Commonly Used Terms

Controller

The *National Electrical Code® (NEC®)** defines a *controller* as a device or group of devices that serves to govern in some predetermined manner the electric power delivered to the apparatus to which it is connected (*Article 100. Definitions*).

Control

Control, as applied to control circuits, is a broad term that means anything from a simple toggle switch to a complex system of components that may include relays, contactors, timers, switches, and indicating lights. One example of a simple control circuit is a light switch used to turn lights on and off. A complex control circuit can accomplish high degrees of precise, automatic machine operation.

Of course, there are many devices and equipment systems in industrial applications. Beyond being used to start or stop a motor, motor control can be used to protect a motor, associated machinery, and personnel. Motor controllers might also be used for reversing, changing speed, jogging, sequencing, or indicating a pilot light.

Control Elements

The *control elements* of a control circuit include all of the equipment and devices concerned with the circuit function. These include enclosures, conductors, relays, contactors, pilot

*National Electrical Code® and NEC® are registered trademarks of the National Fire Protection Association.

devices, and overcurrent protection devices. The selection of control equipment for a specific application requires a thorough understanding of controller operating characteristics and wiring layout. The proper control devices must be selected and integrated into the overall plan.

Note: Article 725 Sections 23, 24, 41, Article 240 Section 3(i), and *Article 430 Sections 71–74* in the *NEC®* covers the application of control circuits. You are encouraged to become familiar with this material.

General Principles of Motor Control*

Purpose of Controller

Objectives to be considered when selecting and installing motor-control components for use with particular machines or systems are described in the following paragraphs.

Starting   Connecting directly across the source of voltage may start the motor. Slow and gradual starting may be required, not only to protect the machine, but also to ensure that the line current inrush on starting is not too great for the power company's system. Some driven machines may be damaged if they are started with a sudden turning effort. The frequency of starting a motor is another factor affecting the controller.

Stopping   Most controllers allow motors to coast to a standstill. Some impose braking action when the machine must stop quickly. Quick stopping is a vital function of the controller for emergency stops. Controllers assist the stopping action by retarding the centrifugal motion of machines and the lowering operations of crane hoists.

Reversing   Controllers are required to change the direction of machine rotation automatically or at the command of an operator at a control station. The reversing action of a controller is a continual process in many industrial applications.

Running   The maintainenance of desired operational speeds and characteristics is a prime function of controllers. Controllers protect motors, operators, machines, and materials while the motors are running. There are many different types of safety circuits and devices to protect people, equipment, and industrial production and processes against possible injury that may occur while the machines are running.

Speed Control   Some controllers can maintain very precise speeds for industrial processes. Other controllers can change the speed of a motor either in steps or gradually through a continuous range of speeds.

Operator Safety   Many mechanical safeguards have been replaced or aided by electric means of protection. Electric-control pilot devices in controllers provide a direct means of protecting machine operators from unsafe conditions.

Protection from Damage   Part of the operation of an automatic machine is to protect the machine itself as well as the manufactured or processed materials it handles. For example, a certain motor-control function may be the prevention of conveyor pileups. A motor control can reverse, stop, slow down, or do whatever else is necessary to protect the machine or processed materials.

*Adapted with permission from Herman, *Industrial Motor Control 4,* Delmar Publishers.

Maintenance of Starting Requirements: Once properly installed and adjusted, motor starters provide reliable operation of starting time, voltages, current, and torque for the benefit of the driven machine and the power system. The *NEC®,* supplemented by local codes, governs the selection of the proper sizes of conductors, starting fuses, circuit breakers, and disconnect switches for specific system requirements.

Manual Control

Manual control is accomplished by mechanical means. The effort required to actuate the mechanism is almost always provided by a human operator. The motor may be manually controlled using any one of the following devices.

Toggle Switch A *toggle switch* is a manually operated electric switch. Many small motors are started with toggle switches. This means the motor may be started directly without the use of magnetic switches or auxiliary equipment. Motors started with toggle switches are protected by the branch-circuit fuse or circuit breaker. These motors generally drive fans, blowers, or other light loads.

Safety Switch In some cases it is permissible to start a motor directly across the full line voltage if an externally operated *safety switch* is used. The motor receives starting and running protection from dual-element, time-delay fuses. The use of a safety switch requires manual operation. A safety switch, therefore, has the same limitations common to most manual starters.

Drum Controller *Drum controllers* are rotary, manual switching devices that are often used to reverse motors and to control the speed of AC and DC motors. They are used particularly where frequent start, stop, or reverse operation is required. These controllers may be used without other control components in small motors, generally those with fractional horsepower ratings. Drum controllers are used with magnetic starters in large motors.

Remote and Automatic Control

A motor may be controlled by remote control using push buttons. When push-button remote control is used or when automatic devices do not have the electric capacity to carry the motor starting and running currents, magnetic switches must be included. Magnetic-switch control is accomplished by electromagnetic means. The effort required to actuate the electromagnet is supplied by electric energy rather than by the human operator. If the motor is to be automatically controlled, the following two-wire pilot devices may be used.

Float Switch The raising or lowering of a float that is mechanically attached to electric contacts may start motor-driven pumps to empty or fill tanks. *Float switches* are also used to open or close piping solenoid valves to control fluids.

Pressure Switch *Pressure switches* are used to control the pressure of liquids and gases (including air) within a desired range. Air compressors, for example, are started directly or indirectly on a call for more air by a pressure switch.

Time Clock *Time clocks* can be used when a definite *on-and-off period* is required and adjustments are not necessary for long periods of time. A typical requirement is a motor that must start every morning and shut off every night at the same time.

Thermostat   In addition to pilot devices sensitive to liquid levels, gas pressures, and time of day, *thermostats* sensitive to temperature changes are widely used. Thermostats indirectly control large motors in air-conditioning systems and in many industrial applications to maintain the desired temperature range of air, gases, liquids, or solids. There are many types of thermostats and temperature-actuated switches.

Limit Switch   *Limit switches* are designed to pass an electric signal only when a predetermined limit is reached. The limit may be a specific position for a machine part or a piece of work, or a certain rotating speed. These switches take the place of a human operator and are often used under conditions where it would be impossible or impractical for the operator to be present or to efficiently direct the machine.

Limit switches are used most frequently as overtravel stops for machines, equipment, and products in process. These devices are used in the control circuits of magnetic starters to govern the starting, stopping, or reversing of electric motors.

Electric or Mechanical Interlock and Sequence Control   Many of the electric control devices described in this chapter can be connected in an interlocking system so that the final operation of one or more motors depends upon the electric position of each individual control device. For example, a float switch may call for more liquid but will not be satisfied until the prior approval of a pressure switch or time clock is obtained. To design, install, and maintain electric controls in any electric or mechanical interlocking system, the electrician must understand the total operational system and the function of the individual components. With practice, it is possible to transfer knowledge of circuits and descriptions to an understanding of additional, similar controls. It is impossible in instructional materials to show all possible combinations of an interlocking control system. However, by understanding the basic functions of control components and their basic circuitry and by taking the time to trace and draw circuit diagrams, the electrician can begin to understand difficult interlocking control systems.

Starting and Stopping

In starting and stopping a motor and its associated machinery, there are a number of conditions that may affect the motor. A few of them are discussed here.

Frequency of Starting and Stopping   The starting duty cycle of a controller is an important factor in determining how satisfactorily the controller will perform in a particular application. Magnetic switches such as motor starters, relays, and contactors, actually beat themselves apart from opening and closing thousands of times. An experienced electrician soon learns to look for this type of component failure when troubleshooting any inoperative control panels. National Electrical Manufacturers Association (NEMA) standards require that the starter size be derated if the frequency of start-stop, jogging, or plugging is more than five times per minute. Therefore, when the frequency of starting the controller is great, the use of heavy-duty controllers and accessories should be considered. For standard-duty controllers, more frequent inspection and maintenance schedules should be followed.

Light- or Heavy-Duty Starting   Some motors may be started with no loads, and others must be started with heavy loads. When motors are started, large feeder-line disturbances may be created that can affect the electric distribution system of the entire industrial plant. The disturbances may even affect the power company's system. As a result, power companies and electric-inspection agencies place certain limitations on *across-the-line* motor starting.

Fast or Slow Start (Hard or Soft)   To obtain the maximum twisting effort (torque) of the rotor of an AC motor, the best starting condition is to apply full voltage to the motor terminals. The driven machinery, however, may be damaged by the sudden surge of motion. To prevent this type of damage to machines, equipment, and processed materials, some controllers are designed to start slowly and then increase the motor speed gradually in definite steps. Power companies and inspection agencies often use this type of starting to avoid electric line surges.

Smooth Starting   Although reduced electric and mechanical surges can be obtained with a step-by-step motor-starting method, very smooth and gradual starting requires different controlling methods. These are discussed in detail later in the text.

Manual or Automatic Starting and Stopping   While the manual starting and stopping of machines by an operator is still a common practice, many machines and industrial processes are started and restarted automatically. The use of automatic devices results in tremendous savings of time and materials. Automatic stopping devices are used in motor-control systems for the same reasons. Automatic stopping devices greatly reduce the safety hazards of operating some types of machinery, both for the operator and for the materials being processed. A brake may be required to stop a machine's motion in a hurry to protect materials being processed or people in the area.

Quick Stop or Slow Stop   Many motors are allowed to coast to a standstill. However, manufacturing requirements and safety considerations often make it necessary to bring machines to a stop as rapidly as possible. Automatic controls can retard and brake the speed of a motor and also apply a torque in the opposite direction of rotation to bring about a rapid stop. This is referred to as *plugging*. Plugging can only be used if the driven machine and its load will not be damaged by the reversal of the motor torque. The control of deceleration is one of the most important functions of a motor control.

Dynamic Braking   Another method of braking electric motors is known as *dynamic braking*. When this method is used to reduce the speed of DC motors, the armature is connected across a load resistor when power is disconnected from the motor. If the field winding of the motor remains energized, the motor becomes a generator and current is supplied to the load resistor by the armature. The current flowing through the armature winding creates a magnetic field around the armature. This magnetic field causes the armature to be attracted to the magnetic field of the pole pieces. This action in a DC generator is known as *counter torque*. Thus, using countertorque to brake a DC motor is known as dynamic braking.

Alternating current induction motors can be braked by momentarily connecting DC voltage to the stator winding (Figure 1–1). When DC is applied to the stator winding of an AC motor, the stator poles become electromagnets. Current is induced into the windings of the rotor as the rotor continues to spin through the magnetic field. This induced current produces a magnetic field around the rotor. The magnetic field of the rotor is attracted to the magnetic field produced in the stator. The attraction of these two magnetic fields produces a braking action in the motor.

An advantage of using dynamic braking is that motors can be stopped rapidly without wearing brake linings or drums. It cannot be used to hold a suspended load, however. Mechanical brakes must be employed when a load must be held, such as with a crane or hoist.

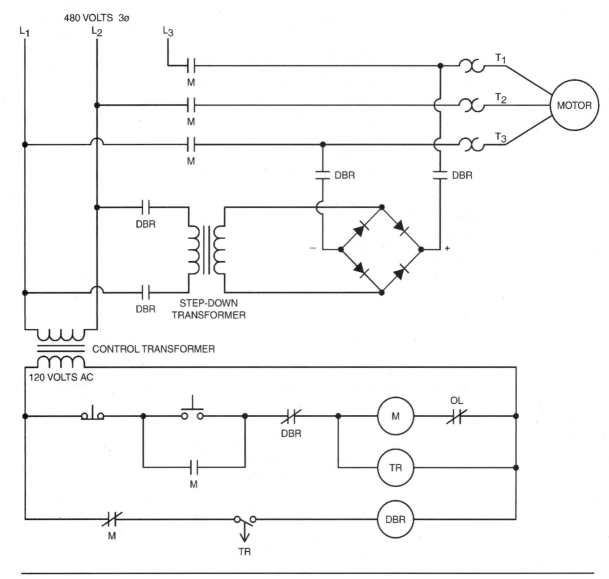

Figure 1–1 Dynamic braking for an AC motor.

Accurate Stops An elevator must stop at precisely the right location so that it is aligned with the floor level. Such accurate stops are possible with the use of automatic devices interlocked with control systems.

Frequency of Reversal Frequent reversal of the direction of motor rotation imposes large demands on the controller and the electric-distribution system. Special motors and special starting and running protective devices may be required to meet the conditions of frequent reversal. A heavy-duty drum switch-controller is often used for this purpose.

Speed Control of Motors

The *speed control* is concerned not only with starting the motor but also with maintaining or controlling the motor speed while it is running. There are a number of conditions to be considered for speed control.

Constant Speed   Constant-speed motors are used on water pumps. Maintenance of constant speed is essential for motor generator sets under all load conditions. Constant-speed motors with ratings as low as 80 rpm and horsepower ratings up to 5000 HP are used in direct-drive units. The simplest method of changing speeds is by gearing. Using gears, almost any predetermined speed may be developed by coupling the input gear to the shaft of a squirrel-cage induction motor.

Varying Speed   A varying speed is usually preferred for cranes and hoists. In this type of application, the motor speed slows as the load increases and speeds up as the load decreases.

Adjustable Speed   With adjustable speed controls, an operator can gradually adjust the speed of a motor over a wide range while the motor is running. The speed may be preset, but, once it is adjusted, it remains essentially constant at any load within the rating of the motor.

Multispeed   For multispeed motors, such as the type used on turret lathes in a machine shop, the speed can be set at two or more definite rates. Once the motor is set at a definite speed, the speed will remain practically constant regardless of load changes.

Protective Features

The particular application of each motor and control installation must be considered to determine what protective features are required to be installed and maintained.

Overload Protection   *Running protection* and *overload protection* refer to the same thing. This protection may be either an integral part of the motor or separate. A controller with electric-overload protection will protect a motor from burning up while allowing the motor to achieve its maximum available power under a range of overload and temperature conditions. An electric overload on the motor may be caused by mechanical overload on driven machinery; a low line voltage; an open electric line in a polyphase system, resulting in single-phase operation; or motor problems such as too badly worn bearings, loose terminal connections, or poor ventilation within the motor.

Open-Field Protection   Direct current shunt and compound-wound motors can be protected against the loss of field excitation by field-loss relays. Other protective arrangements are used with starting equipment for DC and AC synchronous motors. Some sizes of DC motors may race dangerously with the loss of field excitation, while other motors may not race at all due to friction and the fact that they are small.

Open-Phase Protection   A blown fuse, an open connection, a broken line, or other reasons may cause phase failure in a three-phase circuit. If phase failure occurs when the motor is at a standstill during attempts to start, the stator currents will rise to a very high value and will remain there, but the motor will remain stationary (not turn). Since the windings are not properly ventilated while the motor is stationary, the heat produced by the high currents may damage them. Dangerous conditions are also possible while the motor is running. When the motor is running and an open-phase condition occurs, the motor may continue to run. The torque will decrease, possibly to the point of motor *stall*; this condition is called *breakdown torque.*

Reversed-Phase Protection   If two phases of the supply of a three-phase induction motor are interchanged (phase reversal), the motor will reverse its direction of rotation. In elevator operation and industrial applications, this reversal can result in serious damage. Phase-failure and phase-reversal relays are safety devices used to protect motors, machines, and personnel from the hazards of open-phase or reversed-phase conditions.

Overtravel Protection   Control devices are used in magnetic starter circuits to govern the starting, stopping, and reversal of electric motors. These devices can be used to control regular machine operation, or they can be used as safety emergency switches to prevent the improper functioning of machinery.

Overspeed Protection   Excessive motor speeds can damage a driven machine, materials in the industrial process, or the motor. Overspeed safety protection is provided in control equipment for paper and printing plants, steel mills, processing plants, and the textile industry.

Reversed-Current Protection   Accidental reversal of currents in DC controllers can have serious effects. Direct current controllers used with three-phase, AC systems that experience phase failures and phase reversals are also subject to damage. Reversed-current protection is an important provision for battery charging and electroplating equipment.

Mechanical Protection   An enclosure may increase the *life span* and contribute to the trouble-free operation of a motor and controller. Enclosures with particular ratings such as *general-purpose, watertight, dustproof, explosionproof,* and *corrosion-resistant* are used for specific applications. All enclosures must meet the requirements of national and local electrical codes and building codes.

Short-Circuit Protection   For large motors with greater-than-fractional horsepower ratings, short-circuit and ground-fault protection is generally installed in the same enclosure as the motor-disconnect means. Overcurrent devices (such as fuses and circuit breakers) are used to protect the motor branch-circuit conductors, the motor-control apparatus, and the motor itself against sustained overcurrent due to short circuits, grounds, and prolonged and excessive starting currents.

Classification of Automatic Motor-Starting Control Systems

The numerous types of automatic starting and control systems are grouped into the following classifications: current-limiting acceleration and time-delay acceleration.

Current-Limiting Acceleration

Current-limiting acceleration is also called *compensating time.* It refers to the amount of current or voltage drop required to open and close magnetic switches when used in a motor-accelerating controller. The rise and fall of the current or voltage determine a timing period that is used mainly for DC motor control. Examples of types of current-limiting acceleration are:

- Counter-electromotive force (cemF) or voltage-drop acceleration
- Lockout contact or series-relay acceleration

Time-Delay Acceleration

For this classification, *definite-time relays* are used to obtain a preset timing period. Once the period is preset, it does not vary regardless of current or voltage changes occurring during motor acceleration. The following timers and timing systems are used for motor acceleration; some are also used in interlocking circuits for automatic control systems:

- Pneumatic timing
- Motor-driven timers
- Capacitor timing
- Electronic timers
- Dashpot timers

Trouble-shooting

One of the primary jobs of an industrial electrician is troubleshooting control circuits. An electrician that is proficient in troubleshooting is sought after by most of industry. The greatest troubleshooting tool an electrician can possess is the ability to read and understand control schematic diagrams. Many of the circuits shown in this text are accompanied by detailed explanations of the operation of the circuit. If students study the circuits and explanations step-by-step, they will have an excellent understanding of control schematics when they have completed this text.

Most electricians follow a set procedure when troubleshooting a circuit. If the problem has occurred several times in the past and was caused by the same component each time, most electricians check that component first. If that component proves to be the problem, the electrician has saved much time by not having to trace the entire circuit.

Another method of troubleshooting a circuit is *shotgun troubleshooting*. This method derives its name from the manner in which components are tested. Instead of following the circuit, the electrician quickly checks the major components of the circuit. This approach is used to save time because in many industrial situations an inoperative piece of equipment can cost a company thousands of dollars for each hour it is not working.

When neither of these methods reveals the problem, the electrician must use the control schematic to trace the circuit in a logical step-by-step procedure. The primary tool used to trace a circuit is the volt-ohmmeter (VOM), which measures voltage, current, and resistance. It is often necessary to use jumper leads to bridge open contacts when using the VOM. When a jumper lead is used for this application, it should be provided with short-circuit protection. This can be done by connecting a small fuse holder or circuit breaker in series with the jumper lead. In this way, if the jumper is accidentally shorted, the fuse or circuit breaker will open and protect the rest of the circuit.

When troubleshooting a circuit, most electricians work backward through the circuit. For example, one line of a control schematic is shown in Figure 1–2. The relay coil M is connected in series with a normally closed overload contact, a normally open limit-switch contact, a normally closed pressure-switch contact, a normally closed control relay (CR) contact, and a normally open float-switch contact. The problem with the circuit is that relay coil M will not energize. The first test should be to measure the voltage at each end of the circuit to confirm the presence of control voltage. The next procedure is to connect the voltmeter across each of the circuit components to determine which one is open and to stop the current flow to the coil. When the voltmeter is connected across a closed contact, there is no voltage drop and the meter indicates 0 volts. If the voltmeter is connected across an open contact, the meter indicates the full voltage of the circuit.

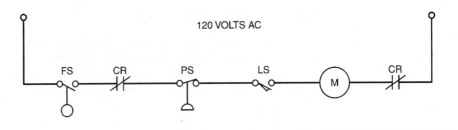

Figure 1–2   Troubleshooting a circuit.

Assume in this circuit that the full circuit voltage is indicated when the meter is connected across float switch *FS*. This reading signals that float switch *FS* is open. The next step is to determine if the switch is bad or if the liquid lever it is sensing has not risen high enough to close the switch. Once that has been determined, the electrician can correct the problem.

Methods of Motor Control

**Manual
Control**

Manual control occurs when someone must initiate an action for the circuit to operate. For example, someone might have to flip the switch of a manual starter to start and stop a motor.

Basic Operation

A manual starter is a switch mechanism. A single-pole manual motor starter is shown in Figure 2–1. Each set of contacts is called a *pole.* A starter with two sets of contacts would be called a two-pole starter. When the switch is in the *OFF* (down) position, the contacts are open, preventing current flow to the motor from the power source. When the switch is in the *ON* (up) position, the contacts are closed, allowing current to flow from the power source to the motor.

Starters are connected between the power source and the load. For example, a two-pole or single-phase motor starter is connected to a motor (Figure 2–2). When the switch is in the *OFF* position, the contacts are open, preventing current flow to the motor from the power source. When the switch is in the *ON* position, the contacts are closed, allowing current to flow from the power source L_1 through the motor, returning to the power source L_2. This is also represented in a schematic drawing, as illustrated in Figure 2-3.

Overloads

Current flow in a conductor always generates heat due to resistance. The greater the current flow, the hotter the conductor. Excessive heat is damaging to electric components. For that reason, conductors have a rated continuous current-carrying capacity or ampacity. Overcurrent protection devices are used to protect conductors from excessive current flow. Thermal overload relays are designed to protect the conductors (windings) in a motor. These protective devices are designed to keep the flow of current in a circuit at a safe level to prevent the circuit conductors from overheating.

Excessive current is referred to as *overcurrent.* The *NEC*® defines overcurrent as any current in excess of the rated current of equipment or the ampacity of a conductor. It may result from overload, short circuit, or ground fault (*Article 100. Definitions*).

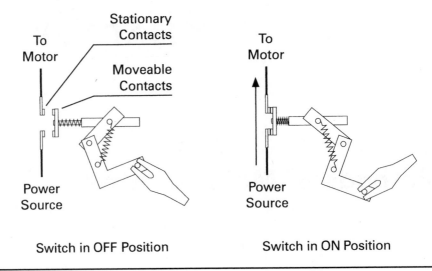

Figure 2–1 Basic operation of a manual starter. Courtesy Siemens Energy and Automation, Inc.

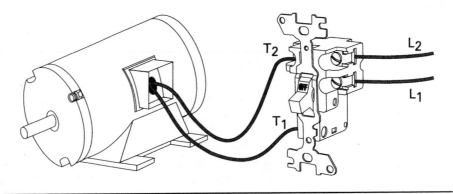

Figure 2–2 Two-pole manual starter. Courtesy Siemens Energy and Automation, Inc.

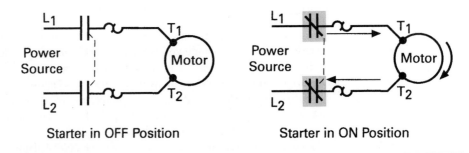

Figure 2–3 Schematic of a two-pole manual starter. Courtesy Siemens Energy and Automation, Inc.

An *overload* occurs when too many devices are operated on a single circuit or when a piece of electric equipment is made to work harder than its design permits. For example, a motor rated for 10 amperes may draw 20, 30, or more amperes in an overload condition. A conveyor can become jammed, causing the motor to work harder and draw more current. Because the motor is drawing more current, it heats up. Damage will occur to the motor in a short time if the problem is not corrected or if the circuit is not shut down by the overload relay.

One method of providing overload protection with a manual starter is with a melting alloy overload relay (see Figure 2–4). The overload relay is connected in series between the line and the motor. Current flows from L_1 through a normally closed (NC) contact and a heater element to the motor. During normal operation, the melting alloy is solid and holds the ratchet wheel firmly in place. The contacts are held closed by a pawl that is connected to the ratchet wheel.

If an overload occurs, the current flow through the heater element is sufficient to cause the melting alloy to melt. The ratchet wheel is free to turn. Spring tension pushes the reset button up, opening the normally closed contacts and stopping the motor. Current flow no longer flows through the heater element, which allows the melting alloy to solidify. Once the melting alloy has become solid, depressing the reset button can close the contacts.

In a motor circuit with a manual starter and overloads, current flows through the overloads while the motor is running. Excessive current flow will cause the overload elements to heat up and, at a predetermined level, open the circuit between the power source and the motor. After a predetermined amount of time, the overload elements will have cooled sufficiently to allow the starter to be reset. When the cause of the overload has been identified and corrected, the motor can be restarted. Some fractional horsepower starters provide overload protection and manual *ON/OFF* control for small motors. Starters are available in one- or two-pole versions suitable for AC motors up to one horsepower and 277 volts AC. The two-pole version is suitable for DC motors up to ¾ horsepower and 230 volts DC. Manual starters are available in a variety of enclosures. A two-speed version is available.

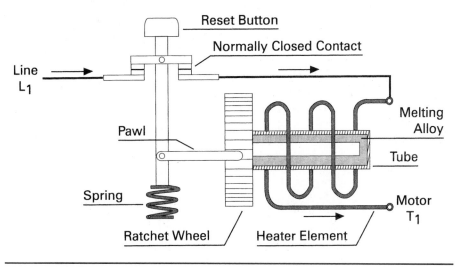

Figure 2–4   Mechanical operation of an overload (OL). Courtesy Siemens Energy and Automation, Inc.

Manual switches are similar to the manual starters but do not provide overload protection. Manual switches only provide manual *ON/OFF* control of single- or three-phase AC motors where overload protection is provided separately. These devices are suitable for use with three-phase AC motors up to 10 horsepower and 600 volts AC and up to 1½ horsepower and 230 volts DC.

The overload relay on some manual starters has adjustable overload ranges. Overload protection is accomplished with the use of a thermal overload with a bimetal strip. This component consists of a small heater wired in series with the motor and a bimetal strip that is linked to the switch's trip mechanism.

A bimetal strip is made of two dissimilar metals bonded together. The two metals have different thermal expansion characteristics, so the bimetal bends at a given rate when heated. As current rises, heat also rises. When a motor overload causes excessive current flow, the resultant rising heat from the heater element causes the bimetal to bend. The hotter the bimetal becomes, the more it bends, until the mechanism is released. The bending action moves the trip lever, opening the main switch contacts.

Automatic Operation

While manual operation of machines is still common practice, many machines are started and stopped automatically. Frequently, there is a combination of manual and automatic control. A process may have to be started manually but is stopped automatically.

Handoff Automatic Switches

Handoff automatic switches are used to select the function of a motor controller either manually or automatically. This selector switch may be a separate unit or may be built into the starter enclosure cover. A typical control circuit using a single-break selector switch is shown in Figure 2–5.

With the switch turned to the *HAND* (manual) position, coil *M* is energized all the time and the motor runs continuously. In the *OFF* position, the motor does not run at all. In the *AUTO* position, the motor runs whenever the two-wire control device is closed. An operator does not need to be present. The control device may be a pressure switch, a limit switch, a thermostat, or another two-wire control pilot device.

Multiple Push-Button Stations

The conventional three-wire, push-button control circuit may be extended by the addition of one or more push-button control stations. The motor may be started or stopped from a number of separate stations by connecting all start buttons in parallel and all stop buttons

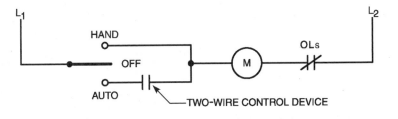

Figure 2–5   Standard-duty, three-position selector switch in control circuit.

in series. The operation of each station is the same as that of the single push-button control in the basic three-wire circuit. Note in Figure 2–6 that pressing any stop button de-energizes the coil. Pressing any start button energizes the coil.

When a motor must be started and stopped from more than one location, any number of start and stop buttons may be connected. Another possible arrangement is to use only one start-stop station and several stop buttons at different locations to serve as emergency stops. Multiple push-button stations are used to control conveyor motors on large shipping and receiving freight docks.

Interlocking Methods for Reversing Control

Interchanging any two motor leads to the line can reverse the direction of rotation of three-phase motors. If magnetic control devices are to be used, then reversing starters accomplish the reversal of the motor direction. Reversing starters wired to NEMA standards interchange lines L_1 and L_3 (Figure 2–7A). To accomplish this, two contactors are needed, one for forward direction and the other for the reverse direction (Figure 2–7B). A technique called *interlocking* is used to prevent the contactors from being energized simultaneously or closing together and causing a short circuit. There are three basic methods of interlocking.

Mechanical Interlock

A *mechanical interlock* is assembled at the factory between the forward and reverse contactors. This interlock locks out one contactor at the beginning of the stroke of either contactor to prevent short circuits and burnouts.

The mechanical interlock between the contactors is represented by the broken line between the coils in Figure 2–8. The broken line indicates that coils *F* and *R* cannot close contacts simultaneously because of the mechanical interlocking action of the device.

When the forward contactor coil *F* is energized and closed through the forward push button, the mechanical interlock prevents the accidental closing of coil *R*. Starter *F* is blocked by coil *R* in the same manner. The first coil to close moves a lever to a position that prevents the other coil from closing its contacts when it is energized. If an oversight allows the second coil to remain energized without closing its contacts, the excess current in the coil due to the lack of the proper inductive reactance will damage the coil.

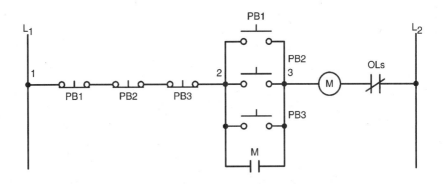

Figure 2–6   Three-wire control using a momentary contact, multiple push-button station.

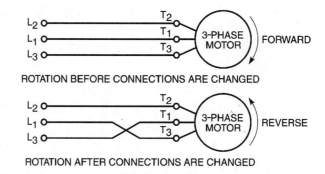

ROTATION BEFORE CONNECTIONS ARE CHANGED

ROTATION AFTER CONNECTIONS ARE CHANGED

Figure 2–7A Reversing rotation of an induction motor.

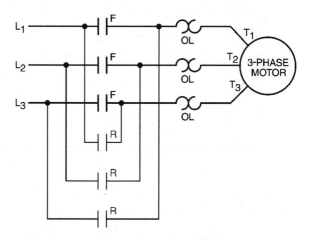

Figure 2–7B Elementary diagram of a reversing starter power circuit.

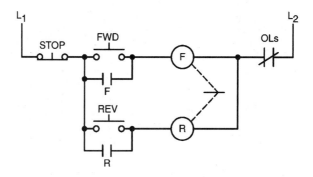

Figure 2–8 Mechanical interlock between the coils prevents the starter from closing all contacts simultaneously. Only one contact can close at a time.

A mechanical interlock is installed on the majority of reversing starters, which are available in horizontal and vertical construction.

Push-Button Interlock

A *push-button interlock* is an *electric method* of preventing both starter coils from being energized simultaneously. When the forward button in Figure 2–9A is pressed, coil *F* is energized and the normally open (NO) contact *F* closes to hold in the forward contactor. Because the normally closed (NC) contacts are used in the forward and reverse push-button units, there is no need to press the stop button before changing the direction of rotation. If the reverse button is pressed while the motor is running in the forward direction, the forward control circuit is deenergized and the reverse contactor is energized and held closed.

Repeated reversals of the direction of motor rotation are not recommended. Such reversals may cause the overload relays and starting fuses to overheat; this disconnects the motor from the circuit and the driven machine may be damaged. It may be necessary to wait until the motor has coasted to a standstill.

NEMA specifications call for a starter to be derated. That is, the next-size-larger starter must be selected when it is to be used for *plugging* to stop or *reversing* at a rate of more than five times per minute. Reversing starters consisting of mechanical and electric interlocked devices are preferred for maximum safety.

Auxiliary Contact Interlock

Another method of electric interlock consists of normally closed *auxiliary contacts* on the forward and reverse contactors of a reversing starter (Figure 2–9B). When the motor is running forward, a normally closed contact *F* on the forward contactor opens and prevents the reverse contactor from being energized by mistake and closing. The same operation occurs if the motor is running in reverse.

The term *interlock* is also used generally when referring to motor controllers and control stations that are interconnected to provide control of production operations. To reverse the direction of rotation of single-phase motors, either the starting or running winding motor leads are interchanged, but not both. Figure 2–10(A) completes the wiring diagram for the single-phase, four-wire, split-phase induction motor; Figure 2–10(B) is a wiring diagram for a single-phase vertical starter; and Figure 2–10(C) is a line diagram of the connections.

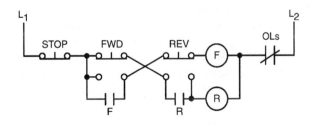

Figure 2–9A Double-circuit push buttons are used for push-button interlocking.

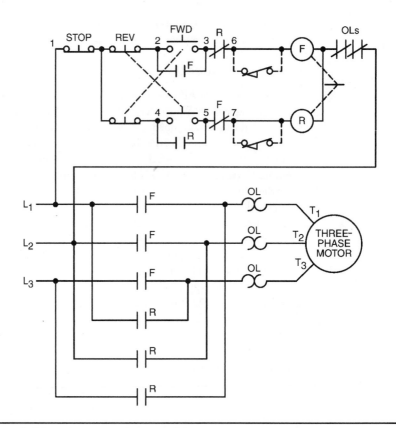

Figure 2–9B Elementary diagram of the reversing starter. The mechanical push-button and auxiliary contact interlocks are indicated.

Methods of Control

Methods of control involve motion, pressure, temperature, time, and counting. In the simplest form of control, the machine function reaches its preset position or condition from only the original or preset values. The circuitry for this application is accomplished with a closed- or an open-loop system. For example, in Figure 2–11, a hydraulically powered piston is to be at position *A* for start conditions. On the operation of a *CYCLE START* pushbutton switch, the piston is to move to position *B,* stop, and return to position *A.* A limit switch is used to control both positions. In many cases, this type of control is accurate enough to satisfy a given production requirement. However, there are many potential problems involving things such as friction, oil temperatures in hydraulic systems, valve-shifting time, material being processed, and many others. All of these conditions may lead to errors that cannot be tolerated in a given production machine. It then becomes necessary to account for the errors and adopt some method to use them. This may be accomplished by using closed-loop control.

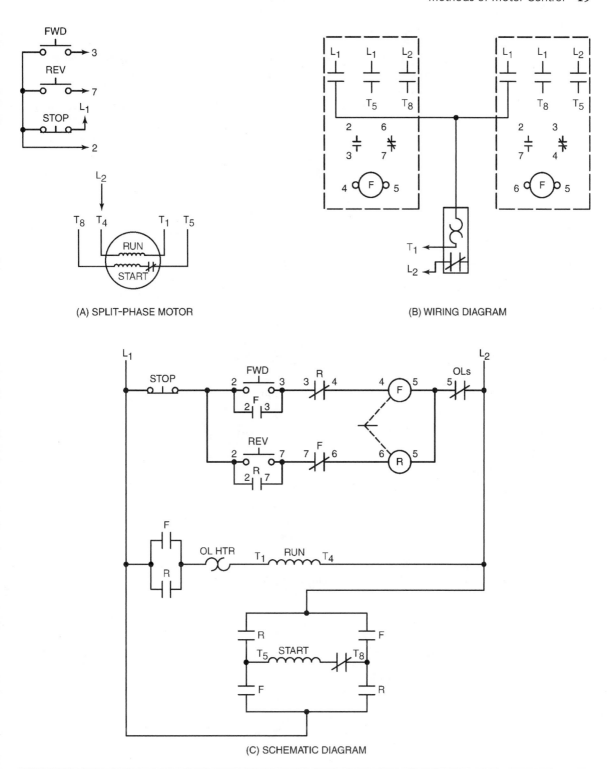

Figure 2–10   Sizes 0 and 1 reversing starters used with single split-phase induction motors.

Closed-Loop Control

A typical *closed-loop control* consists of an amplifier, a servovalve, a cylinder, and position transducers (Figure 2–12). Transducers for the various types of control are covered in subsequent chapters.

In the closed-loop circuit shown in Figure 2–12, the load is to be moved to a desired position. An input signal is fed into the amplifier, which in turn feeds a signal to the servovalve. The valve creates an oil flow to the cylinder that moves the load. To close the loop,

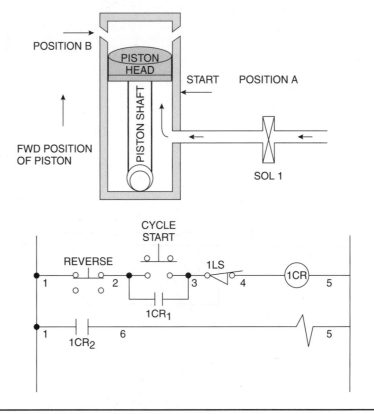

Figure 2–11   Control circuit for hydraulic piston.

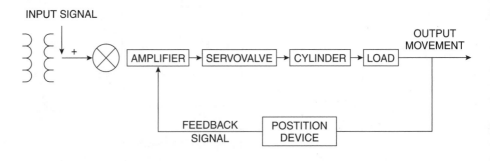

Figure 2–12   Closed-loop control circuit.

a position transducer is connected to the load and provides a feedback signal proportional to the load position. The amplifier compares the required load position (input signal) with the actual load position (feedback signal) and produces an output (control signal) proportional to the difference:

Control signal = (input signal – feedback signal) × gain

Proportional Control

The *control signal* can be thought of as the error between the input and the output. While an error exists, the servovalve will operate to move the load. When the load reaches the desired position, the error signal will be zero. The servovalve centers, and the load stops moving. This type of control is known as *proportional control* because the control signal is proportional to the difference between the input (demand) and the feedback (actual) signals.

In examining the response of such a system (a step change in input signal), there will be a time delay between applying the input signal and the load reaching the desired position. Obviously, the load current cannot move instantaneously to follow the input signal (Figure 2–12), so there will be a time delay between applying the input signal and the load reaching the desired position.

During the first part of the load movement, the system is overcoming the inertia of the load and accelerating it to the required speed. There may then be a period of constant speed movement in which the servovalve is wide open and the system is moving the load as fast as it is able. As the load approaches the required position, the speed reduces (curve flattens out) and the load gradually moves to the demand position. The flattening out at the end is due to the fact that the control signal to the servovalve is basically the error between the input and the output; as the error decreases, the control signal decreases, thus slowing down the movement of the load.

It may be possible to speed up the response of the system by increasing the gain of the amplifier so that for a given error the amplifier produces a larger control signal. Although the output arrives at the desired position faster, it may now overshoot the mark since the load is traveling so fast it cannot be stopped quickly enough when it reaches the required position. There may in fact be several overshoots and undershoots before it settles down to the final position. In many applications, an overshoot may be undesirable, so an alternative method of improving the system response is to modify the control signal so that it is proportional not only to the error but also to the rate at which the error is changing (error rate):

Control signal = (error signal × gain) + (rate of change of error × gain)

When the error is decreasing, for example, the load is moving toward the required position; the error-rate term will then be negative and tend to decrease the control signal. Mathematically, the rate of change of a quantity is known as the *derivative*, so this type of control is termed *proportional plus derivative*.

Control signal = [proportion signal (error) × proportional gain] +
[derivative signal (error rate) × derivative gain]

Consider now a situation in which the load has to be positioned in such a way that the load is subjected to some external force. Imagine that the load has moved to its required position. When actual position and demand position are the same, the error will be zero, and the servovalve will center. However, due to the load on the piston and leakage across the servovalve spool, the piston will tend to creep down. To counteract this, therefore, the servovalve spool must be displaced slightly away from center to prevent movement of the load. To offset the servovalve spool, a small control signal is required. In order for there to be a control signal, an error must exist between input and output; that is, the load can never be in exactly the required position. This is known as the *steady-state error.*

Proportional, Integral, Derivative: The steady-state error can be reduced by again modifying the control signal so that it is now proportional to the error plus an error × time term. The error × time component is known as the *integral term* and is effectively equal to the error multiplied by the length of time the error has existed. Thus, the longer the time, the larger the term becomes. This can therefore be added to the proportional signal to virtually eliminate the steady-state error:

$$\text{Control signal} = [\text{proportional signal (error)} \times \text{proportional gain}] + [\text{integral signal (error time)} \times \text{integral gain}]$$

If a steady-state error exists, the integral term will increase with time until it is large enough to generate the control signal required to correct the error. Adjustment of the proportional and integral gains allows the system to be tuned for the best results. In some applications, it may be necessary to reset the integral term to zero each time a new demand signal is applied to the system.

The performance of a basic (proportional) control system can therefore be significantly improved by adding in:

- A *derivative (error-rate) term* to improve the dynamic response and reduce overshoot
- An *integral (error-time) term* to reduce the steady-state error.

In fast-acting systems, the integral term tends to make the system less stable, so it is often used in conjunction with the derivative term, which tends to improve the stability. The system would then be described as having *proportional, integral, and derivative (PID) action.* This is sometimes referred to as *three-term control.*

Importance of Position Indication and Control

In the subject of electric control for machines, position plays an important part. The problem is to accurately and reliably supply position information by providing an adequate electric signal. This information may be used for both indication and control. Indication becomes a powerful tool in troubleshooting a machine or process.

In many cases, the relative position of a machine part or process product may not be too critical. However, in some cases, position information must be reliable to 0.001 inch or less. The machine designer must know the operation of the machine and/or the product process. The degree of accuracy can then be determined. There are many components that are used to obtain position information, and most are covered in this chapter.

Motion Control

Jogging (Inching)*

Jogging describes the frequent starting and stopping of a motor for short periods of time. A motor would be *jogged* when a piece of driven equipment has to be positioned fairly closely, for example, when positioning the table of a horizontal boring mill during setup. If jogging

*Adapted with permission from Square D Company.

is to occur more frequently than five times per minute, NEMA standards require that the starter be derated. A NEMA type-1 starter has a normal duty rating of 7.5 horsepower at 230 volts, polyphase. On jogging applications, this same starter has a maximum rating of 3 horsepower.

Jogging Control Circuits*   The control circuits covered in this unit are representative of the various methods that are used to obtain jogging. Figure 2–13 is a line diagram of a very simple jogging control circuit. The stop button is held open mechanically. With the stop button held open, maintaining contact *M* cannot hold the coil energized after the start button is closed. The disadvantage of a circuit connected in this manner is the loss of the lock-stop safety feature. This circuit can be mistaken for a conventional three-wire control circuit locked off for safety reasons, such as to keep a circuit or machine from being energized. A padlock should be installed for safety purposes. If the lock-stop push button is used for jogging, it should be clearly marked for this purpose.

Figure 2–14 illustrates other simple schemes for jogging circuits. The normally closed push-button contacts on the jog button in the bottom illustration are connected in series with the holding circuit contact on the magnetic starter. When the jog button is pressed, the normally open contacts energize the starter magnet. At the same time, the normally closed contacts disconnect the holding circuit. When the button is released, therefore, the starter immediately opens to disconnect the motor from the line. The action is similar in the top illustration. A jogging attachment can be used to prevent the reclosing of the normally closed contacts of the jog button. This device assures that the starter holding circuit is not reestablished if the jog button is released too rapidly. Jogging can be repeated by reclosing the jog button; it can be continued until the jogging attachment is removed.

CAUTION: If the circuits shown in Figure 2–14 are used without the jogging attachment mentioned, they are hazardous. A control station using such a circuit, less a jogging attachment, can maintain the circuit when the operator's finger is quickly removed from the button. This could injure production workers, equipment, and machinery. Responsible people committed to safety in the electric industry should not use this circuit.

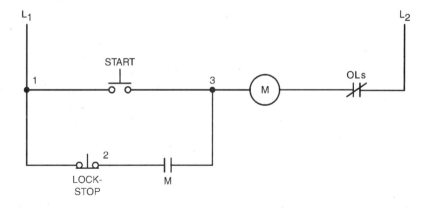

Figure 2–13   Lock-stop push button in jogging circuit.

*Adapted with permission from Herman, *Industrial Motor Controls,* Delmar Publishers.

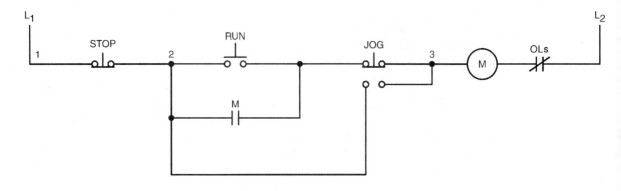

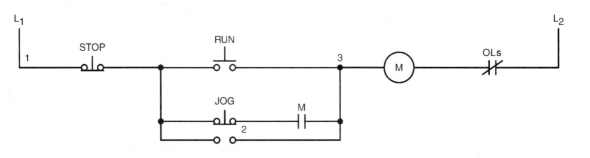

Figure 2–14   Line diagrams of simple jogging control circuits.

Jogging Using a Control Relay   When a jogging circuit is used, the starter can be energized only as long as the jog button is depressed. This means the machine operator has instantaneous control of the motor drive.

The addition of a *control relay* to a jogging circuit provides even greater control of the motor drive. A control-relay jogging circuit is shown in Figure 2–15. When the start button is pressed, the control relay is energized and a holding circuit is formed for the control relay and the starter magnet. The motor will now run. The jog button is connected to form a circuit to the starter magnet. This circuit is independent of the control relay. As a result, the jog button can be pressed to obtain the jogging or inching action.

Other typical jogging circuits using control relays are shown in Figure 2–16. In Figure 2–16(A), pressing the start button energizes the control relay. In turn, the relay energizes the starter coil. The normally open starter interlock and relay contact then form a holding circuit around the start button. When the jog button is pressed, the starter coil is energized independently of the relay and a holding circuit does not form. As a result, a jogging action can be obtained.

Jogging with a Control Relay on a Reversing Starter   The control circuit shown in Figure 2–17A permits the motor to be jogged in either the forward or the reverse direction while the motor is at standstill or is rotating in either direction. Pressing either the start-forward button or the start-reverse button causes the corresponding starter coil to be energized. The

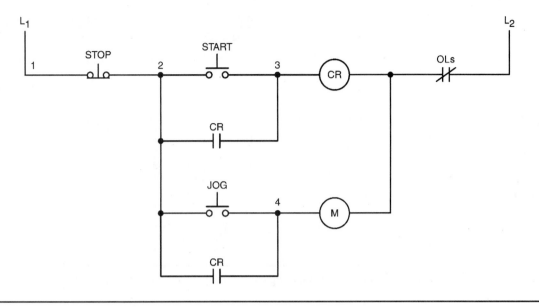

Figure 2–15 Jogging is achieved with added use of control relay.

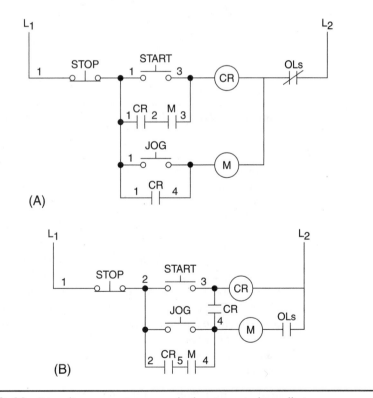

Figure 2–16 Line diagrams using control relays in typical installations.

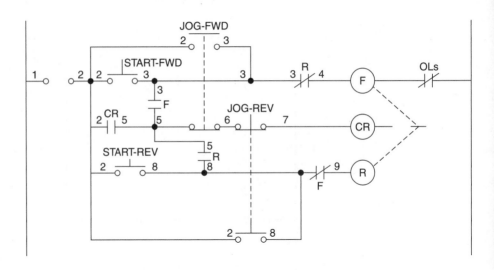

Figure 2–17A Jogging using control relay on a reversing starter.

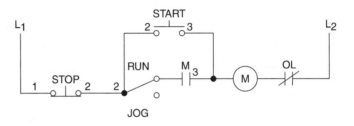

Figure 2–17B Jogging using a standard-duty, two-position selector switch.

coil then closes the circuit to the control relay, which picks up and completes the holding circuit around the start button. While the relay is energized, either the forward or the reverse starter will also remain energized. If either jog button is pressed, the relay is deenergized and the closed starter is released. Continued pressing of either jog button results in a jogging action in the desired direction.

Jogging with a Selector Switch   The use of a selector switch in the control circuit to obtain jogging requires a three-element control station with start and stop controls and a selector switch. A standard-duty, two-position selector switch is shown connected in the circuit in Figure 2–17B. The starter maintaining circuit is disconnected when the selector switch is placed in the jog position. The motor is then inched with the start button.

The use of a selector push button to obtain jogging is shown in Figure 2–18. In the jog position, the holding circuit is broken and jogging is accomplished by depressing the push button.

Jogging with a Push-Pull Operator   Another type of jog-run control can be connected using a push-pull operator. The push-pull operator used in this circuit contains two normally open momentary contacts (Figure 2–19). When the control is pulled outward, contact *A* completes a circuit to coil *CR*. When coil *CR* energizes, both *CR* contacts close.

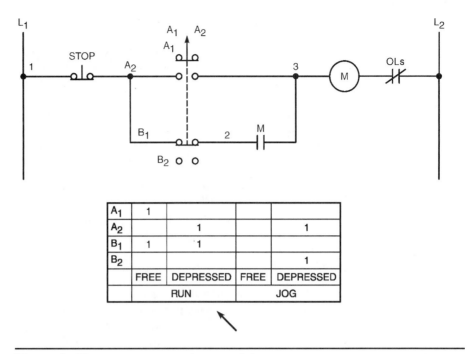

A_1	1			
A_2		1		1
B_1	1	1		
B_2				1
	FREE	DEPRESSED	FREE	DEPRESSED
		RUN		JOG

Figure 2-18   Jogging using a selector push button.

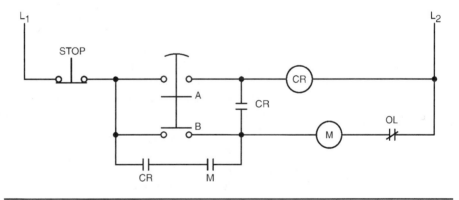

Figure 2-19   Jog-run control using a push-pull operator.

Contact CR_1 completes a circuit to the coil of motor starter M. When motor starter M energizes, contact M closes. Since contacts CR_2 and M are closed, a circuit is maintained to coil M when the push-pull operator is released and movable contact A returns to its normally open position. The circuit will remain in this condition until the stop button is pushed and coils CR and M deenergize.

When the push-pull operator is pressed, movable contact B completes a circuit to coil M. When coil M energizes, auxiliary contact M closes. Contact CR, however, is open and there is no complete circuit to maintain current flow to coil M. When the push button is released and movable contact B returns to its open position, the circuit to coil M is broken and the motor starter deenergizes.

Plugging*

When a motor running in one direction is momentarily reconnected to reverse the direction, it will be brought to rest very rapidly. This is referred to as *plugging*. If a motor is plugged more than five times per minute, derating of the controller is necessary due to the heating of the contacts.

Plugging can only be used if the driven machine and its load will not be damaged by the reversal of the motor. Any standard reversing controller can be plugged, either manually or with electromagnetic controls. Before the plugging operation is attempted, however, several factors must be considered, including:

1. The need to determine if methods of limiting the maximum permissible currents are necessary, especially with repeated operations and DC motors

2. The need to examine the driven machine to ensure that repeated plugging will not cause damage to the machine

Plugging Switches and Applications   *Plugging switches,* or zero-speed switches, are designed to be added to control circuits as pilot devices to provide quick, automatic stopping of machines. In most cases, squirrel-cage motors will drive the machines. If the switches are adjusted properly, they will prevent the direction reversal of rotation of the controlled drive after it reaches a standstill following the reversal of the motor connections. One typical use of plugging switches is for machine tools that must stop suddenly at some point in their cycle of operation to prevent inaccuracies in the work or damage to the machine. Another use is for processes in which the machine must stop completely before the next step of work begins. In this case, the reduced stopping time means that more time can be applied to production to achieve a greater total output. The shaft of a plugging switch is connected mechanically to the motor shaft or to a shaft on the driven machine. The rotating motion of the motor is transmitted to the plugging-switch contacts either by a centrifugal mechanism or by a magnetic induction arrangement (eddy current disk) within the switch. The switch contacts are wired to the reversing starter that controls the motor. The switch acts as a link between the motor and the reversing starter. The starter applies just enough power in the reverse direction to bring the motor to a quick stop.

Plugging a Motor to a Stop from One Direction Only   The forward rotation of the motor in Figure 2–20 closes the normally open plugging-switch contact. When the stop button is pushed, the forward contactor drops out. At the same time, the reverse contactor is energized through the plugging switch and the normally closed forward interlock. Thus, the motor connections are reversed and the motor is braked to a stop. When the motor is stopped, the plugging switch opens to disconnect the reverse contactor. This contactor is used only to stop the motor using the plugging operation; it is not used to run the motor in reverse.

Adjustment   The torque that operates the plugging-switch contacts will vary according to the speed of the motor. An adjustable contact spring is used to oppose the torque to ensure that the contacts open and close at the proper time regardless of the motor speed. To operate the contacts, the motor must produce a torque that will overcome the spring pressure. The spring adjustment is generally made with screws that are readily accessible when the switch cover is removed.

---

*Adapted with permission from Herman, *Industrial Motor Controls,* Delmar Publishers.

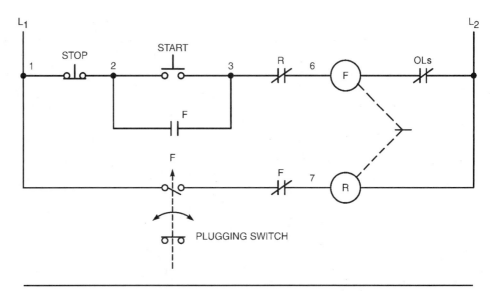

Figure 2–20 Plugging motor to stop from one direction only.

Care must be exercised to prevent the entry of chips, filings, and hardware into the housing when it is opened. Such material may be attracted to the magnets, or they may hamper spring action. The housing must be carefully cleaned before the cover is removed for maintenance or inspection.

Installation To obtain the greatest possible accuracy in braking, the switch should be driven from the shaft with the highest available speed that is within the operating range of the switch. Gears may drive the plugging switch by a chain drive or by a direct flexible coupling. The preferred method of driving the switch is to connect a direct flexible coupling to a suitable shaft on the driven machine. The coupling must be flexible since the centerline of the motor or machine shaft and the centerline of the plugging-switch shaft are difficult to align accurately enough to use a rigid coupling. The switch must be driven by a positive means. Thus, a belt drive should not be used. In addition, a positive drive must be used between the various parts of the machine being controlled, especially where these parts have large amounts of inertia.

The starter used for this type of circuit is a reversing starter that interchanges two of the three motor leads for a three-phase motor, reverses the direction of current through the armature for a DC motor, and reverses the relationship of the running and starting windings for a single-phase motor.

Motor Rotation Experience shows that there is little way of predetermining the direction of the rotation of motors when the phases are connected externally in proper sequence. This is an important consideration for the electrician when the applicable electrical code or specifications require that each phase wire of a distribution system be color coded.

If the shaft end of a motor runs counterclockwise rather than in the desired clockwise direction, the electrician must reconnect the motor leads at the motor. For example, assume that many three-phase motors are to be connected and the direction of rotation of all the motors must be the same. If counterclockwise rotation is desired, the supply phase should be connected to the motor terminals in the proper sequence, T_1, T_2, and T_3. If the motor does not rotate in the desired counterclockwise direction using these connections, the leads may be interchanged at the motor. Once the proper direction of rotation is established, the remaining motors can be connected in a similar manner if they are from the same manufacturer. If the motors are from different manufacturers, they may rotate in different directions even when all the connections are similar and the supply lines have been phased out for the proper phase sequence and color coded. The process of correcting the rotation may be difficult if the motors are located in a place that is difficult to reach.

Lockout Relay   The zero-speed switch can be equipped with a lockout relay or a safety latch relay. This type of relay provides a mechanical means of preventing the switch contacts from closing unless the motor-starting circuit is energized. The safety feature ensures that if the motor shaft is turned accidentally, the plugging-switch contacts do not close and start the motor. The relay coil generally is connected to the T_1 and T_2 terminals of the motor. The lockout relay should be a standard requirement for circuits to protect people, machines, and production processes.

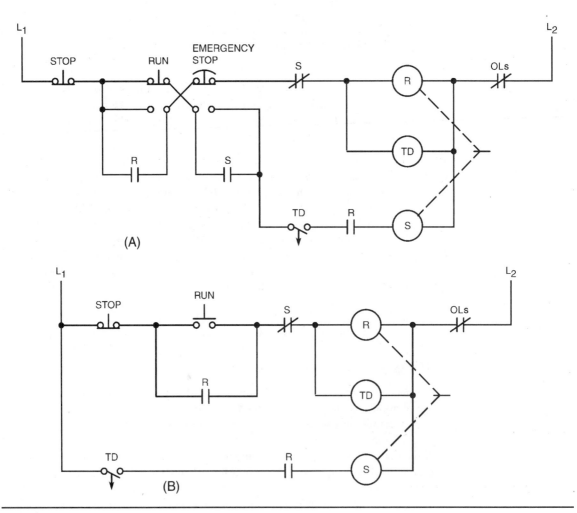

Figure 2–21   Plugging with time-delay relay.

Plugging with the Use of a Timing Relay A time-delay relay may be used in a motor-plugging circuit (Figure 2–21). Unlike the zero-speed switch, this control circuit does not compensate for a change in the load conditions on the motor. The circuit shown in Figure 2–21 can be used for a constant load condition once the timer is preset. If the emergency stop button in Figure 2–21(A) is pushed *momentarily* and the normally open circuit is *not* completed, the motor will coast to a standstill. (This action is also true of the normally closed stop button.) If the emergency stop button is pushed to complete the normally open circuit of the push button, contactor *S* is energized through the closed contacts *TD* and *R*. Contactor *S* closes and reconnects the motor leads, causing a reverse torque to be applied. When the relay coil is deenergized, the opening of contactors can be retarded. The time lag is set so the contact *TD* opens at or near the point at which motor shaft speed reaches 0 rpm.

Alternate Circuits for Plugging Switches The circuit in Figure 2–22 is used for operation in one direction only. When the stop push button is pressed and immediately released, the motor and the driven machine coast to a standstill. If the stop button is held down, the motor is plugged to a stop.

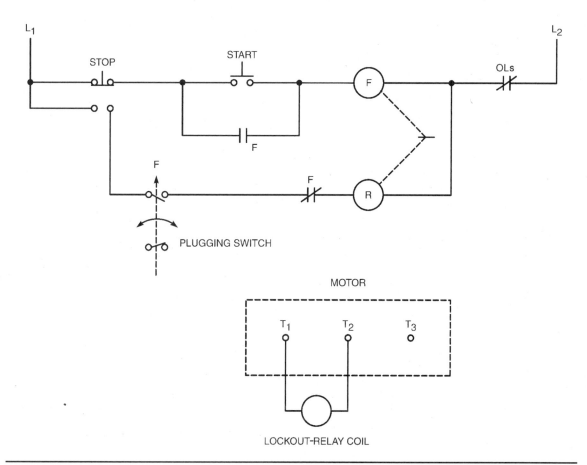

Figure 2–22 Holding stop button stops motor in one direction.

Using the circuit shown in Figure 2–23, the motor may be started in either direction. When the stop button is pressed, the motor can be plugged to a stop from either direction.

The circuit shown in Figure 2–24 provides operation in one direction. The motor is plugged to a stop when the stop button is pressed. Jogging is possible with the use of a control relay.

Antiplugging Protection Antiplugging protection, according to NEMA, is obtained when a device prevents the application of a countertorque until the motor speed is reduced to an acceptable value. An antiplugging circuit is shown in Figure 2–25. With the motor operating in one direction, a contact on the antiplugging switch opens the control circuit of the contactor used to achieve rotation in the opposite direction. This contact will not close until the motor speed is reduced. Then the other contactors can be energized.

Alternate Antiplugging Circuits The motor-starter selector switch controls the direction of rotation of the motor (Figure 2–26). The antiplugging switch completes the reverse circuit only when the motor slows to a safe, preset speed. Under-voltage protection is not available.

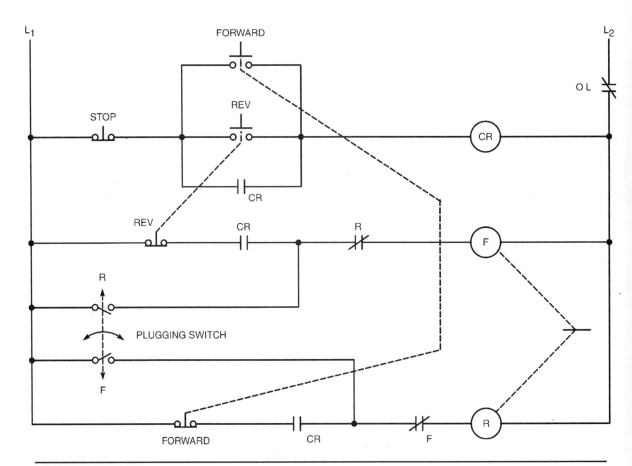

Figure 2–23 In this circuit, pressing stop button stops motor in either direction.

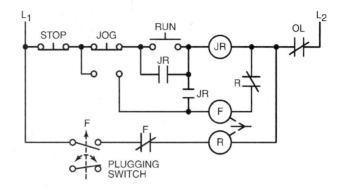

Figure 2–24   Pressing stop button stops motor in one direction.

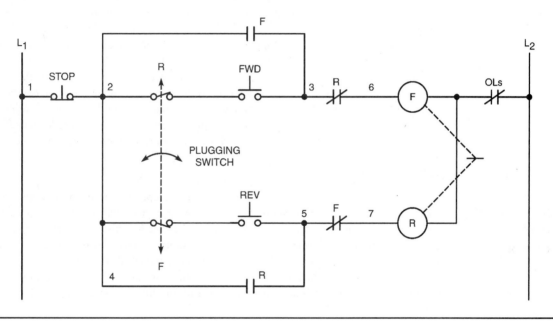

Figure 2–25   Antiplugging protection; the motor is to be reversed but not plugged.

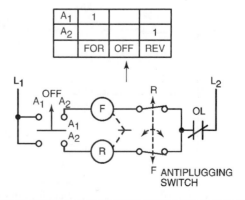

Figure 2–26   Antiplugging with rotation-direction selector switch.

Sequence Control

Many processes require a number of separate motors that must be started and stopped in a definite *sequence,* as in a system of conveyors. When starting up, the delivery conveyor must start first with the other conveyors starting in sequence to avoid a pileup of material. When shutting down, the reverse sequence must be followed with time delays between the shutdowns (except for emergency stops) so that no material is left on the conveyors. This is an example of a simple sequence control.

Sequence control is the method by which starters are connected so that one cannot be started until the other is energized. This type of control is required whenever the auxiliary equipment associated with a machine, such as high-pressure lubricating and hydraulic pumps, must be operating before the machine itself can be operated safely. Another application of sequence control is in main or subassembly line conveyors.

Automatic Sequence Control

A series of motors can be started *automatically* with only one start-stop control station as shown in Figure 2–27 and Figure 2–28. When the lube oil pump, M_1 in Figure 2–28, is started by pressing the start button, the pressure must be built up enough to close the pressure switch before the main drive motor M_2 will start. The pressure switch also energizes a timing relay *TR.* After a preset time delay, the contact *TR* will close and energize the feed-motor starter coil M_3. If the main drive motor M_2 becomes overloaded, the starter and timing relay *TR* will open. As a result, the feed-motor circuit M_3 will be deenergized due to the opening of the contact *TR.* If the lube-oil-pump motor M_1 becomes overloaded, all of the motors will stop. Practically any desired overload control arrangement is possible.

Switch and Circuit Applications for Motion Control

Limit Switches—Mechanical

Mechanical-type switches can be subdivided into those operated from linear motion and those operated from rotary motion. There are large and small switches that operate from linear motion. The precision limit switch is a small switch. This switch varies from the larger size mainly by its lower operating force and shorter stroke. The operating force may be as low as one pound. The stroke may be only a few thousandths of an inch.

Limit switches operated by rotary motion are generally called rotating-cam limit switches. These are control-circuit devices used with machinery having a repetitive cycle of

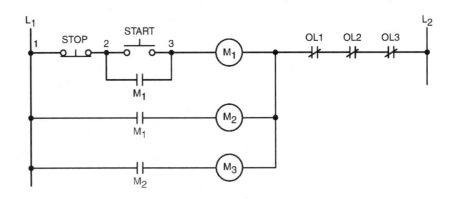

Figure 2–27 Auxiliary contacts (or interlocks) used for automatic sequence control. Contact M_1 energizes coil M_2; contact M_2 energizes coil M_3.

operation in which motion can be correlated to shaft rotation. It is used to limit and control the movement of a rotating machine and to initiate functions at various points in the repetitive cycle of the machine.

The switch assembly consists of one or more snap-action switches. The cams are assembled on a shaft. The shaft in turn is driven either directly or through gearing by a rotary motion on the machine. The cams are independently adjustable for operating at different locations within a complete 360-degree rotation. In some cases, the number of total rotations available is limited. In other cases, the rotation can continue at speeds up to 600 rpm.

In selecting a limit switch, it is important to determine its application in the electric circuit. The following factors must be considered:

- Contact arrangement
- Current rating of the contacts
- Slow or snap action
- Isolated or common connection
- Spring return or maintained
- Number of normally open and normally closed contacts required

In most cases, the switch consists of double-break, snap-action, silver-tipped to solid silver contacts. The contact current rating will vary from 5 amperes to 10 amperes at 120 volts AC continuous. The make-contact rating will be much higher, and the break-contact rating will be lower. Isolated normally open and normally closed contacts are available. In some cases, multiple switches in the same enclosure operated by the same mechanical actions are used.

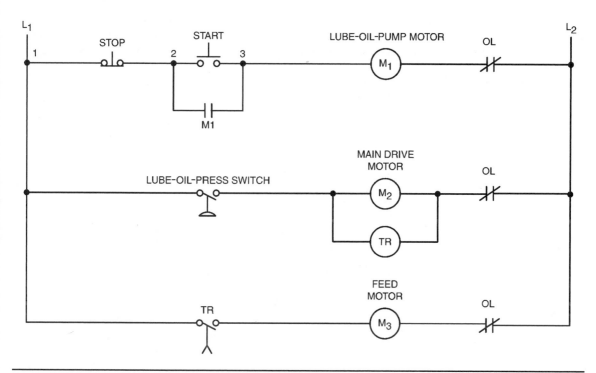

Figure 2–28 Pilot devices used in an automatic sequence control scheme.

A second important factor in selecting a limit switch is the type of mechanical action available to operate the switch. Here selecting the operator is the major decision. Length of travel, speed, force available, accuracy, and type of mounting possible are some of the considerations. In discussing the action of limit switches, several terms are used. A knowledge of these terms is helpful:

1. *Operating force*   The amount of force applied to the switch to cause the *snapover* of the contacts

2. *Release force*   The amount of force still applied to the switch plunger at the instant of *snapback* of the contacts to the unoperated condition

3. *Pretravel or trip travel*   The distance traveled in moving the plunger from its free or unoperated position to the operated position

4. *Overtravel*   The distance beyond operating position to the safe limit of travel, usually expressed as a minimum value

5. *Differential travel*   The actuator travel from the point where the contacts snap over to the point where they snap back

6. *Total travel*   The sum of the trip travel and the overtravel

Figure 2–29 shows the last four items in diagram form.

Most manufacturers of limit switches list this information for certain switches in their specifications. Accuracy of switch operators at the point of snapover varies with different types and manufacturers. In general, it is in the range of 0.001 to 0.005 inch.

The operator that has probably the greatest use is the roller lever. It is available in a variety of lever lengths and roller diameters. The next most frequently used operator is the push rod. It can consist of only a rod, or it can be supplied with a roller in the end. In most cases, particularly with the oil-tight machine-tool limit switch, the head carrying the operator can be rotated to four positions 90 degrees apart. It can also be either top- or side-mounted. Two other operators used in machine control are the fork lever and the wobble stick.

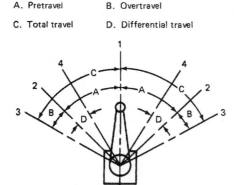

A. Pretravel   B. Overtravel

C. Total travel   D. Differential travel

1. Actuator—free position
2. Actuator—operating position
3. Overtravel—limit position
4. Actuator—release position

LEVER-TYPE SPRING RETURN

PUSH-TYPE SPRING RETURN

Figure 2–29   Limit switch operating movement.

Circuit Applications

Limit switches in machine-tool electric control are used to gather information relative to the position of a machine part. To illustrate this, it is helpful to show a means of providing motion. The valve, along with a cylinder-piston assembly, is used in the limit switch application circuits.

An explanation follows of the solenoid operating valve symbol as it is used and shown in circuit problems (Figure 2–30). Figure 2–30(A) shows the symbol for a single-solenoid, spring-return operating valve in the deenergized condition. Figure 2–30(B) shows the symbol for the same valve with solenoid *A* energized. Note the change in the flow of pressure through the valve to the piston-cylinder assembly.

Figure 2–30(C) shows the symbol for the double-solenoid operating valve with both solenoids deenergized. Figure 2–30(D) shows the symbol with solenoid *A* energized. Figure 2–30(E) shows the symbol with solenoid *B* energized. Here again, as with the single-solenoid operating valve, note the change that takes place when either solenoid is energized. Remember that when either solenoid is deenergized, centering springs in the valve return the piston to the center position.

The limit switch circuit is shown in Figure 2–31. A cylinder-piston assembly is shown as a means of moving a cam on the piston from position *A* to position *B*. If the piston is in position *A*, solenoid *A* is deenergized. If the piston is in position *B*, solenoid *A* is energized. Limit switch *1LS* contact, shown in the circuit of Figure 2–31, is normally closed.

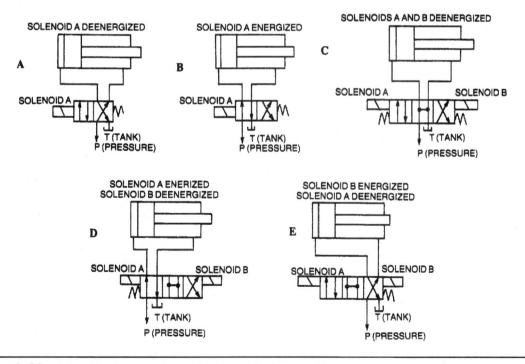

Figure 2–30   Solenoid valve operation with piston-cylinder assembly.

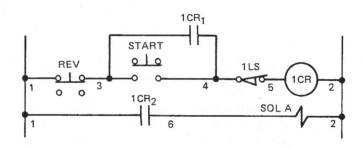

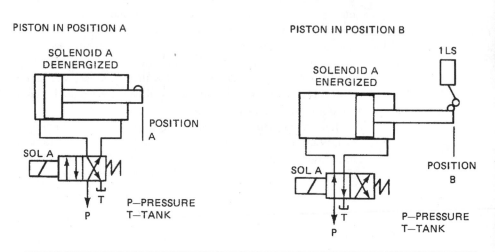

Figure 2–31 Control circuit.

The *sequence of operations* for the circuit shown in Figure 2–31 proceeds as follows:

1. Operate the *START* push-button switch.
2. Relay coil *1CR* is energized.
   (a) Relay contact $1CR_1$ closes, interlocking around the *START* push-button switch.
   (b) Relay contact $1CR_2$ closes, energizing solenoid *A*.
   The valve spool shifts, permitting pressure to enter the main cylinder area (left-hand end). The pressure medium (water, oil, or air) in the right end of the cylinder (rod end) is free to return to the tank. The piston moves from its start position at A to a new position at B.
3. At position *B,* limit switch *1LS* is operated, opening its normally closed contact.
4. Relay coil *1CR* is deenergized.
   (a) Relay contact $1CR_1$ opens, opening the interlock circuit around the *START* push-button switch.
   (b) Relay contact $1CR_2$ opens, deenergizing solenoid *A*.
   With solenoid *A* deenergized, the valve spring returns the valve spool to its initial position. This permits pressure to enter the rod end (right end), returning the piston to position *A*.

In the circuit shown in Figure 2–31, it is assumed that the piston was at position *A* to start. To ensure that the piston is at position *A* to start a cycle, a second limit switch, *2LS*, is placed at position *A* (Figure 2–32). A normally open limit switch contact is used. It is hold operated (contact closed) by a cam on the piston. Note that *2LS* limit switch contact is placed inside the *START* push-button interlock circuit formed by relay contact *$1CR_1$*. Otherwise, as soon as the cam moves off limit switch *2LS,* the limit switch contact opens, deenergizing the circuit.

The operation of this circuit is identical to that shown in Figure 2–31, except the piston must be in position *A,* operating limit switch *2LS* for start conditions.

A third circuit is shown in Figure 2–33. This circuit uses a double-solenoid, spring-return-to-center valve. A limit switch performs a double duty. The normally open contact on limit switch *2LS,* held closed in position *A* (start position), ensures that the piston is at position *A* for start conditions. The normally closed contact on limit switch *2LS* is held open at the start position. This opens the circuit to solenoid *B.*

Proximity Switches

The *proximity switch* consists of a sensor used to detect the presence of an object without physical contact (see Figure 2–34), which is the important difference from the mechanical limit switch. There are several ways to operate the proximity switch. For example, some

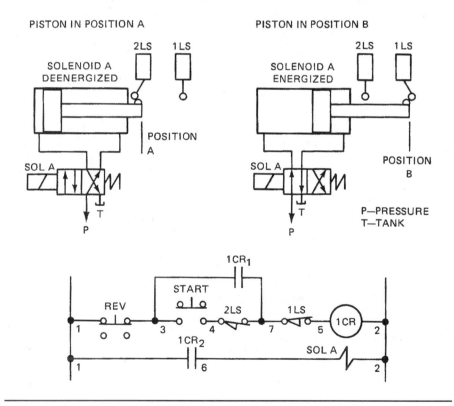

Figure 2–32 Control circuit.

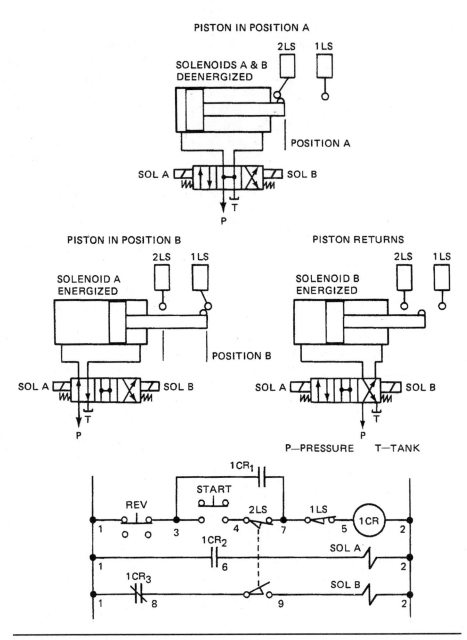

Figure 2–33 Control circuit.

methods use magnetic fields, radio frequency (RF) fields, capacitive fields, acoustic fields, or light rays. *Magnetic fields* have been used to close reed switches by bringing the magnet up close to the switch. Magnets have been used to alter electric fields in devices, making use of the Hall effect.

The *Hall effect* is attributed to E. B. Hall, who first noted the effect late in the last century. The effect is produced when a magnetic field applied to a conductor carrying a current produces a voltage across the conductor. This voltage is known as the *Hall voltage*. The

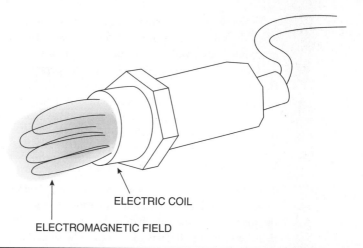

ELECTRIC COIL

ELECTROMAGNETIC FIELD

Figure 2–34 Inductive proximity switch. The proximity switch responds to any object that passes through the electromagnetic field. Inductive proximity switches detect metal objects. Capacitive proximity switches are similar to inductive types but detect nonmetallic and metallic objects.

Hall voltage is proportional to the product of the current and the field. That is, a device that exhibits the Hall effect is a multiplier. With constant current, the Hall voltage is proportional to the magnetic field. With the magnetic field constant, the Hall voltage is proportional to the current flow.

Recently, the production of Hall-effect integrated circuits (1Cs) has eliminated the problems experienced with the use of discrete-component circuit designs. In addition to the use of Hall-effect integrated circuits in proximity switches, designers have obtained good results with the use of capacitive circuits, magnetic pick-up, and photoelectric switching. A word of caution should be added about the use of Hall-effect integrated circuits. They are susceptible to mechanical damage. In hand soldering, a heat sink should be used on the leads.

Radio frequency fields are altered by the presence of ferrous materials. This material absorbs energy by any current produced in it by the field. The *capacitive switch* makes use of a change in capacity that occurs when the object to be sensed acts as a plate of a capacitor. The sensor acts as the other plate. Sonic devices utilize *sound fields* that are either interrupted by the object to be detected or detect the reflection of sound from objects. *Photoelectric devices* operate in a similar manner, but they detect light rays rather than sound waves.

Most proximity switches are supplied in a rugged, molded housing that protects the unit against the dust, oil, and dirt found in industrial installation. Metal housings are also available. For example, stainless steel is used in the food industry. The detection range may vary from 0.1 millimeter to 50 millimeters, with corresponding repeatability of 0.3 millimeter to 5 millimeters.

The *inductive* and *capacitive* types of proximity switches use *oscillators* in their output circuits (see Figure 2–34). The speed of response for these switches is a function of the oscillator frequency. It follows then that the higher the frequency, the shorter the response time. These devices have solid-state outputs. They may have normally open or normally closed output. In some cases, this is programmable.

In many of the switches, a *light-emitting diode* (LED) is supplied. The light goes on when the solid-state output switch is closed. This feature aids the sensitivity adjustment procedure and provides a convenient check of the switch operational status.

The *magnet-operated* proximity switch operates by passing an external magnet near the face of the sensing head. This actuates a small, hermetically sealed reed switch. The 120-volt AC pilot-duty model also includes an epoxy-encapsulated triac output.

One more proximity switch is the *mercury switch*. It operates by passing a permanent magnet of sufficient strength past the switch.

Vane Switches

The *vane switch* is actuated by passage of a separate steel vane through a recessed slot in the switch (see Figure 2–35). Either the vane or switch can be attached to the moving part of the machine. As the vane passes through the slot, it changes the balance of the magnetic field, causing the contacts to operate. The switch is available with either a normally open or normally closed contact and can detect very high speeds of vane travel without detrimental effects such as arm or mechanism wear or breakage. There is no physical contact between the vane and the switch. Therefore, the upper limit on vane speed is governed by factors other than the switch.

The vane switch offers excellent accuracy and response time. Provided that the path of the vane through the slot is constant, repeatability is constant within 10.0025 inches or less. The time required for the switch to operate after the vane has reached the operating point (response time) is less than a millisecond.

Linear-Position Displacement Transducers

In closed-loop control applications, it is necessary to convert some physical property into an electric signal in order to provide a feedback signal to the amplifier. The components that carry out this conversion are known as *transducers*. There are many different devices available to measure linear movement. Factors such as required accuracy and environment aid in the determination of which one to use.

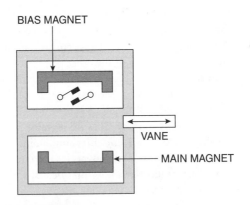

Figure 2–35   Vane switch. When the vane moves inward, it changes the magnetic field, causing the contacts to close.

The simplest device is a *linear potentiometer,* in which the wiper is connected to the moving component and provides a voltage proportional to its position along the potentiometer winding. Problems of linearity, limited resolution, mechanical wear, etc., may occur in practice. The use of linear potentiometers as feedback transducers is therefore rather limited, though they are very inexpensive devices and may be suitable for simpler applications.

To overcome problems of mechanical wear, a noncontact device is required, such as a *linear-variable differential transformer* (LVDT) (see Figure 2–36). It consists of a primary and two secondary coils surrounded by a soft iron core connected to the moving component. The primary coil is connected to a high-frequency AC supply, and voltages are induced in the two secondaries by transformer action. If the two secondary coils are connected in opposition with the core centralized, the induced voltages in each coil cancel out and produce zero output. As the core is moved away from the center, the voltage induced in one secondary increases, while that in the other reduces. This now produces a net output voltage, the magnitude of which is proportional to the amount of movement and the phase determined by the direction. The output can then be fed to a phase-sensitive rectifier (known as a demodulator), which produces a DC signal proportional to movement and polarity, depending on direction.

The *solid-state transducer,* by precisely sensing the position of an external magnet, is able to measure linear displacements with infinite resolution. Since there is no contact between the magnet and the sensor rod, there is no wear, friction, or degradation of accuracy. This transducer's outputs represent an absolute position, rather than an incremental indication of position change. There can be either *digital* or *analog* output.

The measuring principle by which this transducer operates can be explained as follows: When a current pulse is sent through a wire (which has been threaded through a tube and returned outside), the resulting magnetic field is concentrated in the tube, which acts as a wave guide. If this tube is then passed through a doughnut-shaped magnet, the two magnetic fields interact. A tube made of magnetostrictive material will experience a local rotary strain where these fields interact. This strain will continue for the duration of the electrical pulse. The rotary-strain pulse travels along the wave-guide element at ultrasonic speed and can be detected at the end of the tube. By measuring the time from the generation of the

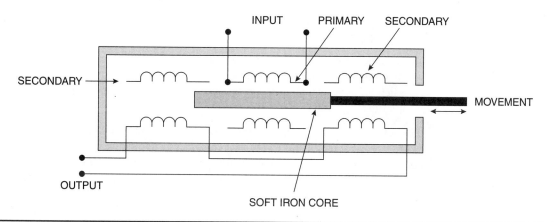

Figure 2–36   Linear-variable differential transformer.

initial electric pulse until the detection of the ultrasonic pulse, one can determine the distance of the external magnet from the reference point.

This transducer is used for either position readout or closed-loop control. It may be mounted inside hydraulic cylinders or externally on machines. The applications are very broad, covering small winding machines, machine tools, plastic forming machines, and so on.

Angular-Position Displacement Transducers

In many industrial applications today, *angular displacement* is an important consideration or requirement. It may range from a few degrees to many revolutions of a shaft. The motion required for a specific application must be accurate in very small increments of change.

The applications of the *angular-position displacement transducer* for motion control are many. They include machining, positioning of tools, testing, inspection, welding, and assembly. In each application, a rotary-variable differential transformer or a drive-motor-and-control combination is designed to satisfy the requirements.

As was the case with linear motion, rotary potentiometers can be used for measuring angular position, but the same problems of wear, and so on, apply, so, again, they are confined to relatively simple applications. The rotary equivalent of an LVDT is known as a rotary-variable differential transformer (RVDT).

A specially shaped cam of magnetic material is rotated inside the primary and secondary coils by the input shaft. The profile of the cam determines the amount of magnetic coupling between the primary and secondary; hence, as with the LVDT, an output signal proportional to shaft rotation is provided. The limitation in this case is the maximum angular movement that can be achieved and still produce a linear output, which in practice may be of the order of +60 degrees.

Rotary Encoder

Another approach to position control that involves the transfer of angular position to a usable signal is the *rotary encoder,* which can be either electromechanical, electronic, optical, or a combination of all three. It can be used to monitor the rotary motion of a device. There are two types:

1. The *incremental encoder,* which transmits a specific quantity of pulses for each revolution of a device.

2. The *absolute encoder,* which provides a specific code for each angular position of the device. It may be in *binary coded decimal* (BCD) or *Gray code.* The Gray code is defined as sequential numbers by binary values in which only one value changes at a time.

The rotary encoder is an optical, incremental, rotary-shaft encoder. It uses an infrared LED and precision optical and mechanical components to produce a series of pulses corresponding to shaft rotation. Output options include standard square wave, directional high/low, and quadrature. It also features reverse polarity and output short-circuit protection.

In summary, then, it can be said that the encoder electronically digitizes shaft motion of a rotating element; that is, it converts mechanical motion to an electronic digital format for a controller. By coupling the encoder shaft to a lead screw, rack and pinion, or rotating element, its electronic output can be used to measure distance from a known position and direction of rotation.

Use of AC Synchronous and DC Stepping Motors

Alternating Current Synchronous Motors

Alternating current synchronous motors are permanent-magnet AC motors with extremely rapid starting, stopping, and reversing characteristics. They have a basic shaft speed of 72 or 200 rpm, depending on the motor selected. Three-lead types need only a single-pole, three-position switch for complete forward, reverse, and OFF control. When operating from a single-phase source, a phase-shifting network consisting of a resistor and a capacitor is used. Motors are available in torque ratings from 22 to 1500 ounce-inches. The output torque varies with changes in the input voltage.

When the motor is energized, AC current flows only through the windings. Current does not flow through the rotor or through brushes, since the motor is of a brushless construction. Therefore, it is not necessary to consider high inrush currents when designing a control for the motor since starting and operating current are, for all practical purposes, identical.

The permanent-magnet construction of the motor provides a small residual torque that holds the motor shaft in position when the motor is deenergized. When holding torque is required, a DC voltage can be applied to one winding when the AC voltage is removed. If necessary, DC voltage can be applied to both windings with a resulting increase in holding torque (see Figure 2–37).

To determine the AC synchronous motor required for a specific application, the electrician must know the following information:

1. *Motor shaft speed.* Motors are available in shaft speeds of 72 and 200 rpm at 60 hertz. Gearing can be used to obtain other speeds. It is also necessary to consider the effects of gears (if used) on torque and inertial load characteristics of the load.

2. *Load characteristics.* Examples that follow show how to determine the torque and moment of inertia characteristics of the load.

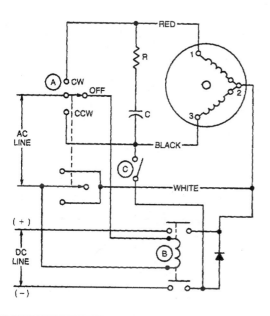

MODE	(A)	(B)	(C)
HOLDING	OFF	CLOSED	CLOSED
START	OPEN	OPEN	CLOSED
	OPEN	OPEN	OPEN
RUN	CW OR CCW	OPEN	OPEN
STOP	OPEN	OPEN	CLOSED
	OFF	CLOSED	CLOSED

SWITCHING SEQUENCE

Figure 2–37 Typical circuit for utilizing DC voltage to provide holding torque.

(a)   *Torque in ounce-inches = Fr*

where F = *force* (in ounces) and r = radius (in inches). See Figure 2–38A.
Using a 4-inch-diameter pulley, it is found that a 2-pound pull on the scale is
required to rotate the pulley:

$$F = 2\ pounds = 32\ ounces$$

The torque is then

$$32 \times 2 = 64\ ounce\text{-}inches$$

(b) Moment of inertia in pound-inches squared (see Figure 2–38B):

$$\frac{I\ (pound\text{-}inches\ squared) = Wr\ squared}{2}\ \text{for a disk}$$

OR

$$\frac{I\ (pound\text{-}inches\ squared) = Wr\ squared}{2}\ \text{for a cylinder}$$

where W = *weight* (in pounds) and r = *radius* (in inches).
Using a load of an 8-inch-diameter gear weighing 8 ounces:

$$W = \frac{8}{16} = 0.5\ pound$$

$$r = \frac{8}{2} = 4\ inches$$

(c) Moment of inertia:

$$\frac{0.5 \times (4) squared}{2} = 4\ pound\text{-}inches\ squared$$

With voltage and frequency known, the user should refer to the manufacturer's tables
of torque and moment of inertia to select the motor that best suits the requirements.

Direct Current Stepping Motor

Another motor, the DC stepping motor, is a permanent-magnet motor that converts electronic signals into mechanical motion. Stepping motors operate on phase-switched DC
power. Each time the direction of current in the motor windings is changed, the motor output shaft rotates a specific angular distance. The motor shaft can be driven in either direction and can be operated at very high stepping rates.

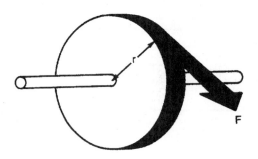

Figure 2–38A   Torque.

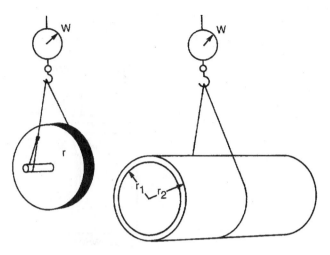

Figure 2–38B Moment of inertia.

The motor shaft advances 200 steps per revolution (1.8 degrees per step) when a four-step input sequence (full-step mode) is used, and 400 steps per revolution (0.9 degrees per step) when an eight-step input sequence (half-step mode) is used. Power transistors connected to flip-flops or other logic devices are normally used for switching, as shown in the wiring diagram in Figure 2–39. The four-step input sequence is also shown. Since current is maintained on the motor windings when the motor is not being stopped, a high holding torque results.

A DC stepping motor offers many advantages as an actuator in a digitally controlled positioning system. It easily interfaces with a microcomputer or microprocessor to provide opening, closing, rotating, reversing, cycling, and highly accurate positioning in a variety of applications.

Mechanical components such as gears, clutches, brakes, and belts are not needed since stepping is accomplished electronically. A DC stepping motor applies holding or detent torque at standstill to help prevent an unwanted motion.

The motors are available in a range of frame sizes and with standard step angles of 0.72 degrees, 1.8 degrees, and 5 degrees. Step accuracies are 3 percent or 5 percent, noncumulative. They can be driven at rates to 20,000 steps per second with a minimum of power input.

In determining the synchronous motor required for a specific DC stepping motor application, the following information must be determined:

- Output speed in steps per second
- Torque in ounce-inches
- Load inertia in pound-inches squared
- Its required step angle
- Time to accelerate in milliseconds
- Time to decelerate in milliseconds
- Type of drive system to be used
- Size and weight considerations

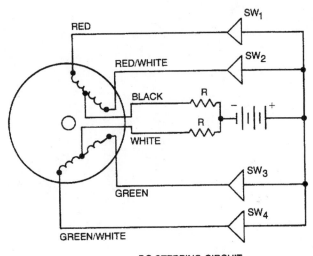

DC STEPPING CIRCUIT

FOUR-STEP INPUT SEQUENCE*
(FULL STEP SEQUENCE)

STEP	SW$_1$	SW$_2$	SW$_3$	SW$_4$
1	ON	OFF	ON	OFF
2	ON	OFF	OFF	ON
3	OFF	ON	OFF	ON
4	OFF	ON	ON	OFF
1	ON	OFF	ON	OFF

* Provides CW rotation as viewed from nameplate end of
motor. To reverse direction of motor rotation perform
switching steps in the following order; 1,4,3,2,1

Figure 2–39 Direct current stepping circuit.

1.   *Torque (ounce-inches) = Fr*

where *F* (in ounces) is force required to drive the load and *r* (in inches) is radius.

2. Moment of inertia:

$$\frac{1\ pound\text{-}inch\ squared = Wr\,2}{2}\ \text{for a disk} =$$

$$\frac{W}{2r1\ squared + r2\ squared}\ \text{for a cylinder}$$

where *W = weight* in pounds and *r = radius* in inches.

3. Equivalent inertia. A motor must be able to:
   (a) Overcome any frictional load in the system
   (b) Start and stop all inertia loads including that of its own rotor.
       The basic rotary relationship is:

$$T = \frac{I \propto}{24}$$

where $T = $ *torque* in ounce-inches
     $I = $ *moment of inertia*
    $\propto = $ *angular acceleration* in radians per second squared
    $24 = $ *conversion factor from ounce-inches per second squared × 24 pound-inches squared*

Measuring angles in radians: A radian is the angle at the center of a circle that embraces an arc equal in length to the length of the radius. The value of the radian in degrees equals 57.296 degrees. Then, 3.1416 radians denotes an angle of 180 degrees. Note that it is often convenient to measure angles in radians when dealing with angular velocity. If $\omega = $ *angular velocity per second of the revolving body* in radians, $V = $ *velocity of a point on the periphery of the body* in feet per second, and $r = $ *radius* in feet, then

$$\frac{V}{r} = \omega$$

For example, if the velocity of a point on the periphery is 10 feet per second and the radius is one foot:

$$\omega = \frac{10}{1} = 10 \; radians \; per \; second$$

Angular acceleration ($\propto$) is a function of the change in velocity (ω) and the time required for the change:

$$\propto = \frac{\omega 2 - \omega 1}{t}$$

Or, if starting from 0:

$$\propto = \frac{\omega}{t}$$

where $\omega = $ *angular velocity* in radians per second and $t = $ *time* in seconds.
Since:

$$\omega = \frac{steps \; per \; second}{steps \; per \; revolution},$$

angular velocity and angular acceleration can also be expressed in steps per second (ω') and steps per second squared ($\propto'$), respectively.

Sample Calculations:
1. To calculate the torque required to rotationally accelerate an inertia load:

$$T = 2 \times Io \left(\frac{\omega}{t} \times \frac{\pi\theta}{180} \times \frac{1}{24} \right)$$

where T = *torque required* in ounce-inches
$\quad Io$ = *inertial load* in pound-inches squared
$\quad \pi$ = *3.1416*
$\quad \theta$ = *step angle* in degrees
$\quad \omega$ = *step rate* in steps per second

Example #1: Assume the following conditions:
- *Inertia = 9.2 pound-inches squared*
- *Step angle = 1.8 degrees*
- *Acceleration = 0 – 1000 steps per second in 0.5 seconds*

$$T = 2 \times 9.2 \times \frac{1000}{.5} \times \frac{1.8}{180} \times \frac{1}{24}$$

T = 48.2 ounce-inches to accelerate inertia

2. To calculate the torque required to accelerate a mass moving horizontally and driven by a rack and pinion or similar device:

The total torque that the motor must provide includes the torque required to:

   (a) Accelerate the weight, including that of the rack
   (b) Accelerate the gear
   (c) Accelerate the motor rotor
   (d) Overcome frictional forces

To calculate the rotational equivalent of the weight: I_{Eq} + *Wr squared* where W = *weight* in pounds and *r = radius* in inches.

Example #2: Assume the following:
- *Weight = 5 pounds*
- *Gear pitch diameter = 3 inches*
- *Gear radius = 1.5 inches*
- *Velocity = 15 feet per second*
- *Time to reach velocity = 0.5 second*
- *Pinion inertia = 4.5 pound-inches squared*
- *Motor rotor inertia = 2.5 pound-inches squared*

$\qquad = Wr2 = 5(1.5)2 = 11.25$ *pound-inches squared*
$\qquad$ *Pinion = 4.5 pound-inches squared*
$\qquad$ *Rotor = 2.5 pound-inches squared*
$\qquad$ *Total = 18.25 pound-inches squared*

Velocity is 15 feet per second, with a 3-inch pitch diameter gear. Therefore:

   (a) Speed = 15 × 12 / 3 × 3.1416 = 19.1 revolutions per second
   (b) The motor step angle is 1.8 degrees (200 steps per revolution).
   (c) Velocity in steps per second: w′ = 19.1 × 200 = 3820 steps per second
   (d) To calculate torque to accelerate the system:

$$T = 2 \times Io \left(\frac{\omega}{t} \times \frac{\pi\theta}{180} \times \frac{1}{24} \right)$$

$= 2 \times 18.25 \times 3820 \times 3.1416 \; 1.8 \times l = 365$ *ounce-inches*

To calculate torque needed to slide the weight, assume a frictional force of 6 ounces. Then

T friction = 6 × 1.5 = 9 ounce-inches

Therefore:

Total torque required = 365 + 9 = 374 ounce-inches

There are many ways in which these motors can be controlled. In general, the design of the control will depend on the particular application.

Photoelectric Transducers

Photoelectric transducers are known as *opto-effectors.* In general, they are divided into four groups (see Figure 2–40A and Figure 2–40B):

1. Through-beam sensors
2. Retroreflective sensors
3. Diffuse-reflection sensors
4. Fiber optic with amplifiers

These transducers are used to detect varying materials at long ranges by noncontact sensing. The transmitters operate with pulsed infrared light (wavelength 880 micrometers using LEDs as the light source. The receivers incorporate a gate circuit (retroreflective and diffuse sensors), thus giving a high immunity to extraneous light, while 20-turn potentiometers above accurate sensitivity setting.

In the *through-beam sensor,* the transmitter and receiver are separately housed and mounted opposite each other. Sensing occurs on the interruption of the beam between the transmitter and the receiver.

The transmitter and receiver in the *retroreflective sensors* are incorporated in a single housing. The beam sent by the transmitter is sent back to the receiver by a reflector. Sensing occurs on the interruption of the beam.

When the *diffuse-reflection sensor* is used, the transmitter and receiver are incorporated in a single housing. The beam sent by the transmitter is reflected back to the receiver by the object to be sensed. The two states, *reflection received* or *no reflection,* are used to determine the presence or absence of an object in the sensing area.

Fiber optics are used for sensing objects in positions where, owing to limited space or high temperatures, it is not possible to mount the transducer. Various sensing heads are available, including miniature designs.

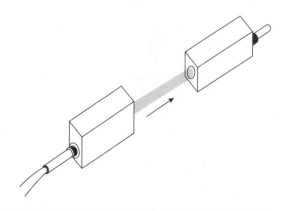

Figure 2–40A Through-beam sensor.

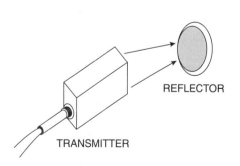

REFLECTOR

TRANSMITTER

Figure 2–40B Retroreflective sensor.

The fiber-optic leads contain light-transmitting glass fibers that are protected by a sheathing. Aluminum-armored units are suitable in hot areas (up to 290° C), and those with polyvinyl chloride (PVC) can be used in areas with temperatures up to 80° C.

Through-beam sensor receivers, retroreflective sensors, and diffuse-reflection sensors have a 20-turn potentiometer for accurate sensitivity setting. The sensitivity should always be set to maximum for through-beam and retroreflective sensors for optimum efficiency, except in the case of transparent objects when it may be necessary to reduce the sensitivity for safe switching. For diffuse-reflection sensors, adjust the sensitivity so that the object to be sensed is recognized without any interference.

The transmitter of a through-beam sensor is fitted with an LED that indicates operating voltage. In the case of through-beam receivers, retroreflective sensors, and diffuse-reflection sensors, the LED is lit when the output closes.

The Automatic Timing and Controls Fiber-Optic System

The *automatic timing and controls fiber-optic system* is an extremely versatile sensing device. Through the use of a variety of reflective or through-beam bundles, this unit can detect objects as small as 0.05 millimeters with a response time of 15 microseconds. It also can detect color marks on backgrounds of different colors. A choice of three LED light sources is available: red or green modulated and red nonmodulated. This unit's output easily interfaces with most programmable logic controllers.

Flow Monitors

The *flow monitor* consists of a sensing head and a control monitor. The system works on the principle of thermal conductivity. The temperature-compensated sensing head is inserted into the system to be monitored and is heated up to a few degrees higher than the medium.

If the medium is flowing, the heat generated in the probe is conducted away; that is, the probe is cooled. If the medium is at rest, the sensor heats up again by a few degrees. The temperature in the sensor is constantly measured and compared with that of the medium.

Thermistors are used to convert the temperature differences into electric signals, which are further processed in the control monitor. So, if the medium is flowing, the probe is cooled and the output relay energizes (LED lights), whereas if there is no flow, the output relay will remain deenergized. There is a maximum delay of 30 seconds after the supply voltage has been applied before the unit is ready to operate.

Pressure Control

Importance of Pressure Indication and Control

Pressure indication and control are important in many electric control systems in which air, gas, or a liquid is involved. The control takes several forms. It may be used to start or stop a machine on either rising or falling pressure. It may be necessary to know that pressure is being maintained.

The *pressure switch* is used to transfer information concerning pressure to an electric circuit. The electric switch unit can be a single, normally open, or normally closed switch. It can be both, with a common terminal, or two independent circuits. Usually, it is easier to use a switch with two independent circuits in the design of a control circuit.

Some terms associated with pressure switches are best explained in chart form. Figure 2–41 shows some of the terms we are concerned with and some of the most important points.

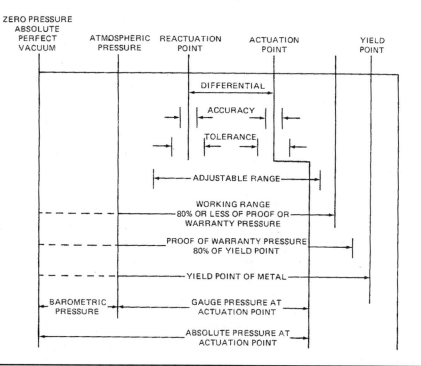

Figure 2–41   Pressure chart.

Tolerance in most switches is about 2 percent; *accuracy* is about one percent. These are expressed in *percent of the working range*. In some cases it is necessary to know and use a *differential pressure*. As used here, the *differential* is the range between the actuation point and the reactuation point. For example, the electric contacts operate at a preset rising pressure and hold operated until the pressure drops to a lower level. This differential may be a fixed amount, generally proportional to the operating range, or it can be adjustable.

The normally closed contact opens upon reaching a preset pressure. The normally open contact closes upon reaching a preset pressure.

Types of Pressure Switches

There are several types of pressure switches available. Some of the most widely used types are:

- Sealed piston
- Bourdon tube
- Diaphragm
- Solid state

Sealed-Piston Pressure Switch   The *sealed-piston* pressure switch is actuated by means of a piston assembly that is in direct contact with the hydraulic fluid. The assembly consists of a piston sealed with an O-ring, direct-acting on a snap-action switch. An extremely long life can be expected from this switch where mechanical pulsations exceed 25 per minute and where high overpressures occur. The sealed piston saves the cost of installing return lines.

The piston-type pressure switch can withstand large pressure changes and high over-pressures. These units are suitable for pressures in the range of 15 pounds per square inch (psi) to 12,000 psi.

Bourdon-Tube Pressure Switch The pressure operating element in the *Bourdon-tube* pressure switch is made up of a tube formed in the shape of an arc. Application of pressure to the tube causes it to straighten out. Care should be taken not to apply pressures beyond the rating of the unit. Excessive pressures tend to bend the tube beyond its ability to return to its original shape.

The Bourdon-tube pressure switch is extremely sensitive. It senses peak pressures; for this reason, it may be necessary to use some form of dampening or snubbing of the pressure entering the tube.

This pressure switch is available for applications from 50 psi to 18,000 psi. It is generally used more in indicating service than in switching. This is due to the inherent low work output of the unit.

Diaphragm Pressure Switch The *diaphragm* pressure switch consists of a disk with convolutions around the edge. The edge of the disk is fixed within the case of the switch. Pressure is applied against the full area of the diaphragm. The center of the diaphragm opposite the pressure side is free to move and operate the snap-action electric contacts. Units are available from vacuum to 150 psi.

Solid-State Pressure Switch The *solid-state* pressure switch is interchangeable with existing electromechanical pressure switches. Pressure sensing is performed by semiconductor strain gauges with proof pressures up to 15,000 psi acceptable. Switching is accomplished with solid-state triacs. An enclosed terminal block allows four different switching configurations:

1. Single-pole, single-throw (normally closed)
2. Single-pole, single-throw (normally open)
3. Single-pole, double-throw
4. Double-make, double-break

Pressure Transducers

With the present industrial use of programmable logic controllers, the use of *pressure transducers* has become important. A pressure transducer utilizes semiconductor strain gauges that are epoxy-bonded to a metal diaphragm. Pressure applied to the diaphragm through the pressure port produces a minute deflection, which introduces strain to the gauges. The strain produces an electric resistance change proportional to the pressure. Four gauges (or two gauges with fixed resistors) form a *Wheatstone bridge*. The differential resistance is measured by applying a constant voltage to the bridge. Diaphragm reflection results in an analog (millivolt) output that is proportional to the pressure. The rigid sensing diaphragm, with bonded semiconductor strain gauges and low mass, provides a no-moving-parts transducer with excellent resistance to shock and vibration.

Circuit Applications

The pressure switch contact is now substituted for the limit switch contact. In Figure 2–42(A) and Figure 2–42(B), the pressure switch is connected into the fluid power pressure circuit. A normally closed contact is connected into the electric circuit (Figure 2–42(C)). Remember that a normally closed pressure switch contact opens on rising pressure.

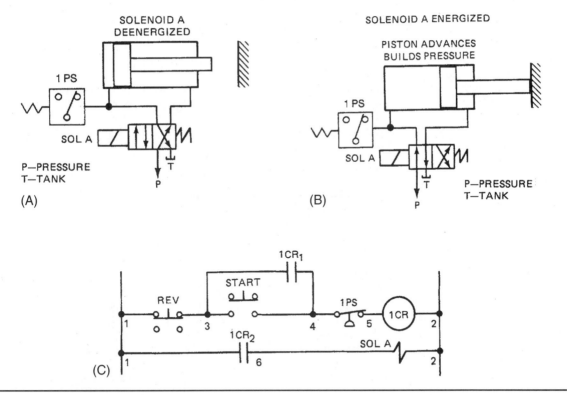

Figure 2–42 Control circuits.

The circuits in Figure 2–43A, Figure 2–43B, Figure 2–43C, and Figure 2–43D use two separate cylinder-piston assemblies. Each assembly is powered through a single-solenoid, spring-return operating valve.

The sequence of operations for these circuits is as follows (see Figure 2–43D):

1. Operate the *START* push-button switch.
2. Relay coil *1CR* is energized.
   (a) Relay contact *1CR$_1$* closes, interlocking around the *START* push-button switch.
   (b) Relay contact *1CR$_2$* closes, energizing solenoid *A*.
3. The piston moves forward (Figure 2–43B).
   On the forward stroke of piston *A*, limit switch *3LS*, normally open contact closes (cam is no longer operating *2LS*). This causes no circuit action as normally closed relay contact *1CR$_3$* is open.
4. The piston reaches the work piece, builds pressure to a preset amount on *1PS*, operating the pressure switch contacts.
5. The normally closed contact on *1LPs* opens, deenergizing relay coil *1CR*.
   (a) Relay contact *1CR$_1$* opens, deenergizing relay coil *1CR*.
   (b) Relay contact *1CR$_2$* opens, deenergizing solenoid *A*.
   (c) Relay contact *1CR$_3$* closes.
      Piston *A* returns to position *A* (Figure 2–43C). On return travel the cam on the piston operates limit switch *3LS*. Contacts *3LS, 2LS, 1CR$_3$* and *2PS* are now closed.

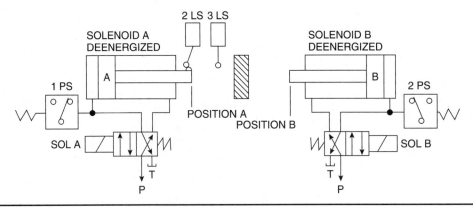

Figure 2–43A A valve-pressure switch–piston-cylinder assembly.

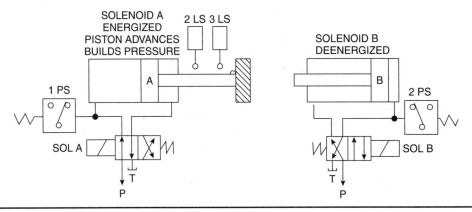

Figure 2–43B A valve-pressure switch–piston-cylinder assembly.

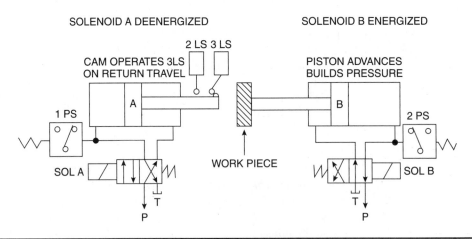

Figure 2–43C A valve-pressure switch–piston-cylinder assembly.

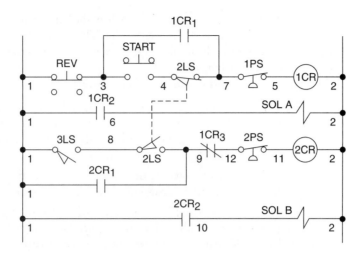

Figure 2–43D Control circuit.

6. Relay coil *2CR* is energized.
   (a) Relay contact $2CR_1$ closes, interlocking around *2LS* normally closed contact and *3LS* normally open contact.
   (b) Relay contact $2CR_2$ closes, energizing solenoid B.
       Piston B now moves forward, meeting the work piece and building pressure to an amount preset on *2PS*.
7. Normally closed *2PS* contact opens, deenergizing relay coil *2CR*.
   (a) Relay contact $2CR_1$ opens, opening the interlock circuit around *3LS* and *2LS*.
   (b) Relay contact $2CR_2$ opens, deenergizing solenoid B.
       Piston B now returns to its initial position at B.

Temperature Indication and Control

The use of *temperature indication and control* is important in at least three areas:

1. Safety

2. Troubleshooting

3. Means of processing material

As a safety issue, it should be noted that excessive heat can cause many problems that lead to unsafe operating conditions, including overheated conductors and excessive heat in operating components, such as circuit breakers and other operating electromechanical components. Many times this is the result of loose connections in a circuit carrying power current. Motor bearings may overheat due to improper lubrication or defective bearings. Overheating in a control cabinet can create false tripping of protective devices.

In processing materials, it is important that the correct temperature be used and maintained. Both indication and control are important for a resulting good product. This is particularly true in the molding of plastics and the die casting of metals. This section is devoted entirely to the use of temperature indication and control in the processing of materials.

Temperature controllers basically consist of two parts: *a sensing element* that responds to temperature and *a switch* consisting of normally open and/or normally closed contacts. The contacts operate from the temperature-sensing element. The operating point is preset. The switch operation is usually accomplished when the temperature at the sensing point changes from any given level to the preset point. The main difference among temperature switches is the means by which the temperature information is transferred from the sensing element to the switching element.

There are several factors to consider when selecting a temperature controller:

1. *Temperature range available. Temperature range* is the overall operating range of the controller. Not all controllers cover the entire temperature range used in industrial control. The controller is generally selected after the range is determined. The sensing element often becomes an important factor.

2. *Type of sensing element.* There are three basic types: 1. electric, 2. differential expansion of metal, and 3. expansion of fluid, gas, or vapor.

3. *Response time.* The response to a temperature change may vary between one type of sensing element and another. Here, the application may be the deciding factor. For example, if rapid response is not required, the slower-response and generally less-expensive types may be used. *Speed of response* is a measure of the elapsed time from the instant the temperature change occurs at the sensing element until it is converted into controller action.

4. *Sensitivity.* Units vary with the amount of temperature change required to operate the switch. This amount is fixed in some switches. In others it is adjustable. It is usually desirable to have a controller that has a relatively high sensitivity. *Resolution sensitivity* is the amount of temperature change necessary to actuate the controller.

5. *Operating differential.* This is the difference in temperature at the sensing element between make and break of the controller's contacts when the controller is cycled.

Electronic Temperature Controller (Pyrometer)

The *electronic controller,* or *pyrometer,* may have one of three different temperature-sensing elements:

- *Thermocouple* (approximate temperature range 200° F to 5000° F)
- *Thermistor* (approximate temperature range 100° F to 600° F)
- *Resistance temperature detector* (RTD) (approximate temperature range 300° F to 1200° F)

The *thermocouple* operates on the principle of joining two dissimilar metals A and B at their extremities (Figure 2–44). When a temperature difference exists between the two extremity points, a potential is generated in proportion to the temperature difference. Such

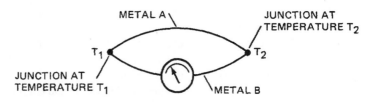

Figure 2–44   Thermocouple junction.

combinations as copper-constantan, iron-constantan, and chromel-alumel are used in thermocouple construction.

When the temperature at T_1 is different than at T_2, electromotive force (emf) exists. By the use of an indicating device that indicates electromotive force or the flow of current, the temperature difference can be determined.

The combination of iron and constantan form the type J thermocouple. The *thermistor* is a semiconductor whose resistance decreases with increasing temperature. This element is connected into a null-reading-type AC bridge circuit with the output fed to an amplifier. The output of the amplifier operates a relay that is used in the control circuit.

The RTD consists of a tube made of stainless steel or brass. It contains an element made in the form of a coil. The coil is generally wound of a fine nickel, platinum, or pure copper wire. As the temperature changes, the resistance of the coil changes proportionally. This change is converted to a voltage by its application in the electric control circuit. Therefore, as the voltage changes with a temperature change, the voltage level is compared to a reference. Thus, an output proportional to the temperature change is obtained.

The electronic temperature controller is usually the most expensive type. It is more sensitive and has a faster response. Generally, electronic controllers have a smaller sensing device. The sensing device can be remotely located with greater ease. Normal control accuracies of one fourth of one percent to five percent of full-scale reading can be expected.

Recent years have brought about rapid advancement and refinement in electronic temperature controllers. Thus, some terms and expressions that may not be too familiar have been developed. Explanations of these terms and expressions are included in the following sections.

On/off is the simplest and oldest method of temperature control. Assuming no time thermal lags in the object being heated or cooled, the on/off control produces a fluctuating temperature. The limits are within the heating or cooling band. Depending on the time lag, the amplitude of the temperature swing may exceed good operating control.

Time proportioning was the next major development in controllers. This method turns the heating elements on and off in accordance with the demands of the temperature set point. There is a point below the set point at which the power to the heating elements is on continuously. Likewise, there is a point above the set point at which the power is off. Between these two points, the power is on at a proportional rate. For example, at a temperature 20 percent through the proportional band, the power may be 80 percent on; 50 percent of the way, it will be 50 percent on. This method differs from on/off control in that it does not require an actual temperature change to cycle.

To accomplish on/off and proportioning, pyrometers can be divided into two types. These are the *millivoltmeter controller* and the *potentiometric controller.*

The *millivoltmeter controller* is the oldest type of pyrometer on the market. It uses a millivoltmeter actuated directly from a temperature sensor such as a thermocouple. Setting the desired temperature positions either a photocell and light source or a set of oscillator coils. The indicating pointer carries a small flag. As the pointer approaches the set point, the flag moves between the photocell and light source or between the oscillator coils. Control action now starts, cutting back on heat.

One of the major sources of trouble with this type of controller is the adverse effect of shock and vibration. Another problem involves the sensing device. For good indication accuracy, the external resistance of the thermocouple circuit should match that of the instrument. Some manufacturers offer controllers with adjustable external resistance circuits. Such instruments can be calibrated after they are installed.

The *potentiometric controller* differs from the millivoltmeter type in that the signal from the temperature sensor is electronically controlled and compared to the set-point temperature. This instrument can be supplied with or without indication because with this control, the indicating method is completely separate.

The potentiometric controller has the following advantages:

- No moving parts
- Does not have to be calibrated for external resistance
- Not affected by shock and vibration

The potentiometric controller has the disadvantage of more electronic circuitry that is not easily serviced; it is generally advisable to replace the entire plug-in controller.

The millivoltmeter and potentiometric controllers both generally use a contactor to energize the heating element load. The contactor is energized and deenergized in response to a demand for heat. For example, when heat is required as sensed by a thermocouple, the contactor is energized by the controller output. Electric power is then connected to the heating element load. When there are no heat requirements, the heating elements are completely isolated from the power source by the opening of the contacts.

Mechanical contactors switch power in full ON and OFF cycles. For this reason they should be used at cycle times of 15 seconds or longer for reasonable service life. Because of the full ON and OFF switching and the limited cycle time, contactor-controlled processes must have a higher tolerance for process temperature overshoot and undershoot.

A more recently developed controller does not use a contactor to energize and deenergize the heating element load. With this latest type of controller, the power remains connected to the load at all times. Just enough power is supplied to the heating elements to satisfy the temperature requirements. This type of controller is called a *stepless* or *proportional controller* and has a current output. It is generally used with silicon-controlled rectifiers (SCR) or triac solid-state power packs. Two different outputs are available with this type of controller. Power is applied to the heating element load through:

- Phase control (phase-angle-fired)
- Zero crossing (zero-angle-fired)

With *phase control,* the output is continuous at the amount required for consistent temperature. This is accomplished by *firing* the SCR at some point through the wave, depending on the amount of power required.

One problem with phase-fired control is the radio frequency interference (RFI) generated. This is caused by the SCR switching at any place in the wave at an extremely fast rate. This is on the order of one-half microsecond (μs). Another problem is the resulting distorted wave shape, which creates a problem for the power company.

These problems can be avoided by the use of zero-crossing SCR circuitry. Its characteristics are similar to the time-proportioning results. However, the power is always switched off and on when the power voltage is zero.

Zero-crossing or *zero-crossover-fired power controllers* are ideal for control of pure resistive loads that can accommodate rapid, full-power, ON/OFF cycling. Zero-crossover firing does not create radio frequency interference (RFI) and will not adversely affect sensitive electronic equipment (computers, other SCR power controllers, logic controllers) located in the same area. Additionally, SCRs are protected from line voltage transients, making them more reliable in a wide variety of applications. Zero-crossover-area SCRs, when coupled with a time-proportioning control (firing package) will operate in a series of full ON and OFF cycles known as *time-proportional burst firing*. The time-proportioning control accepts the control output signal and converts it into a time-proportional signal, determining the amount of ON time and OFF time per duty cycle. The continuous, highly repetitious rate of full ON and full OFF produces a smooth power output to the load (heater) and a stable process temperature. A typical power control system consists of an RTD or thermocouple, a temperature controller, a firing package, and SCR power controller. Often, the firing package is part of the temperature controller and is not a separate component.

These components work together to control the heating of the process:

1. The *temperature sensor* provides a signal to the temperature controller.

2. The *temperature controller* compares the sensor signal to the predetermined set point and generates an output signal that represents the difference between the actual process temperature and the set point.

3. The *firing package* uses this control output to generate a time-proportional signal for the SCR power controller, switching the SCR on and off, thus regulating the power.

Controller Outputs   The relay in pyrometer outputs generally is rated at 5 amperes to 10 amperes. This means it can handle a small heating element load of approximately 500 watts (W) to 1000 watts at 120 volts. For larger loads, the relay is used to energize a contactor.

Controllers are now available with a solid-state output capacity capable of 20 amperes inrush and one ampere continuous. This control output can be used with load contactors up to sizes 2 or 3.

Additional Terms   In discussing the use of pyrometers in the control of temperature, there are several terms that need explanation:

1. *Bandwidth*   This is sometimes referred to as the *proportioning band*. In most controllers, this band is adjustable. If the bandwidth is set for 3 percent, this is 24° F on an 800° F instrument. If the set point is at 450° F, the band extends from 437° F to 463° F. In controlling, it is quite possible for the temperature to settle out at some temperature other than the set point. For example, assume the temperature settles out at 454° F. This represents an offset of +4 degrees. This can be corrected by adjusting the set point to 446° F. Resetting of the control point in this way is called *manual reset*.

2. *Automatic reset*   The addition of the auto reset function in a controller automatically eliminates the offset. Circuitry in the controller recognizes the offset. It will electronically shift the band up or down the scale as required to remove an error. The auto reset feature is sometimes referred to as the *integral function*.

3. *Rate*   The application of rate control is valuable in applications having rapid changes in temperature caused by external heating and cooling. Rate control works in the opposite direction of reset and at a faster speed. For example, a sudden cooling causes the controller to turn the heaters full ON. This is done by an upward shift of the band. Rate is sometimes referred to as the *derivative function*.

4. *Mode*   A two-mode controller is one having proportioning and auto reset, or proportional and auto reset. A three-mode controller is one having proportioning, auto reset, and rate, or proportional, auto reset, and rate.

5. *Analog and digital set point*   With an *analog set point,* the temperature is set on a scale. With the *digital set point* controller, the temperature is displayed.

Temperature Switches (Thermostats)

Temperature switches using differential expansion of metals may be of two different types. One uses a mechanical link; the other uses a fused bimetal. The mechanical link has a temperature range of 100° F to 1500° F. The bimetal type has a temperature range of 40° F to 800° F.

The *mechanical-link-type switch* has one metal piece directly connected or subjected to the part where the temperature is to be detected. The metal expands or contracts due to temperature change. This produces a mechanical action that operates a switch.

The *bimetal-type-switch* operates on the principle of uneven expansion of two different metals when heated. With two metals bonded together, heat will deform one metal more than the other. The mechanical action resulting from a temperature change is used to operate a switch. In bimetal units, the sensing element and the switch are generally enclosed in the same container or are adjacent. The bimetal type is usually the smallest and least expensive type of controller. It is not suited for high-precision work. However, its compact, rugged construction lends it to many uses. Accuracies of one percent to 5 percent of full scale can be expected.

With temperature switches using liquid, gas, or vapor, the sensing elements are generally located remotely from the switch. The temperature ranges for these units are as follows:

- Liquid-filled: 150° F to 2200° F (these may be self-contained)
- Gas-filled: 100° F to 1000° F
- Vapor-filled: 50° F to 700° F

These units operate on the principle that when the temperature increases, expansion of the medium (fluid, gas, or vapor) takes place. Thus, a force is exerted on a device that in turn operates a switch. The liquid-filled type that is self-contained (switch and sensing element together) has a relatively fast response time.

A factor to consider in using units with a remote sensing head is the problem with the tube connecting the sensing bulb with the switching mechanism. The tube is easily damaged by mechanical abuse. Also, the response is slower with long lengths of connecting tube.

The switch contact ratings on temperature switches (thermostats) vary from 5 amperes to 25 amperes at 120 volts. They generally consist of one normally open and one normally closed contact. They may have a common connection point, or they may have isolated contacts.

Circuit Applications

In Figure 2–45, a two-position selector switch is substituted for the two push-button switches. A pyrometer is inserted in the circuit to the contactor coil *10CR.* Note that the symbol for the pyrometer shows only the relay contact in the pyrometer and the thermocouple connections. More details on internal circuit construction can be obtained from the manufacturer of the pyrometer.

When the heat OFF/ON selector switch is operated to the ON position, the circuit is complete to the normally open relay contact. If the temperature of the part being sensed is below the set point on the pyrometer, the relay contact is closed. This completes the circuit to the coil of contactor *10CR.* The coil of contactor *10CR* is energized. Contacts *10CR$_1$, 10CR$_2$* and *10CR$_3$* close, energizing the heating elements at powerline voltage.

CONDUCTORS CARRYING LOAD CURRENT AT LINE VOLTAGE ARE DENOTED BY HEAVY LINES.

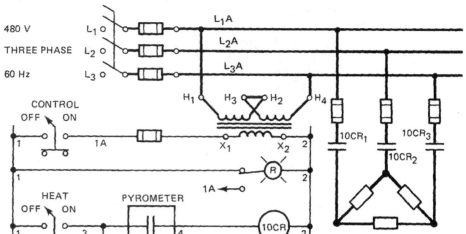

Figure 2–45 Control circuit.

When the temperature of the part being sensed by the thermocouple reaches the set point, the pyrometer-relay contact opens. This deenergizes the coil of contactor *10CR*. Contacts *10CR₁*, *10CR₂*, and *10CR₃* open, deenergizing the heating elements. The pyrometer-relay contact continues to close and open, depending on the thermocouple input. The rate of opening and closing depends on the type of pyrometer.

The temperature switch (thermostat) is generally used at lower temperatures than the pyrometer. It is also used where accuracy and sensitivity are not important factors in temperature control. For example, a tank containing a fluid of some type (oil, asphalt, or brine) is to be heated to a given temperature. Depending on the size of the tank, it may take considerable time to heat it safely to a given temperature. If the preset temperature is 200° F, it generally is not important whether the temperature is actually 204° F or 196° F.

Another use of the thermostat is to indicate and/or control the temperature of fluid used in a machine operation. For example, it may be required that the fluid in a hydraulic system be at a minimum temperature of 70° F for the machine to operate, and the temperature should not exceed 120° F.

The circuits that follow illustrate temperature control of a fluid that is used in a machine for machine operation.

The sequence of operations for the circuit in Figure 2–46 is as follows:

1. Temperature below set point of 125° F, circuit ready to operate.
2. Operate the *START* push-button switch.
3. Energize relay coil *1CR*.
   (a) Relay contact *1CR₁* closes, interlocking around the *START* push-button switch.
   (b) Relay contact *1CR₂* closes, energizing the green pilot light.

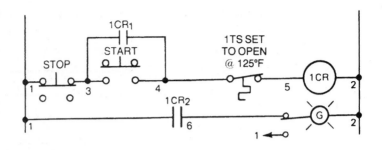

Figure 2–46 Control circuit.

4. Temperature exceeds 125° F, normally closed temperature switch contact opens.
5. Relay coil *1CR* deenergized.
   (a) Relay contact $1CR_1$ opens, opening the interlock circuit around the *START* push-button switch.
   (b) Relay contact $1CR_2$ opens, deenergizing the green pilot light.

The temperature must drop below 125° F, allowing the normally closed *1TS* contact to close before relay coil *1CR* can again be energized.

Figure 2–47 shows the control for two heating elements. With the temperature of the medium being heated (oil, asphalt, brine) below 80° F:

1. Contactor coil *10CR* is energized.
2. Contacts $10CR_1$ and $10CR_2$ close, energizing the two heating elements.
3. Auxiliary contact #2 opens, preventing the energizing of relay coil *1CR* through the operation of the *START* push-button switch.
4. The heated medium now reaches 80° F, deenergizing *10CR* contactor coil by the opening of *1TS*.
5. Contacts $10CR_1$ and $10CR_2$ open, deenergizing the two heating elements.
6. Auxiliary contact closes.
7. The *START* push-button switch can now be operated, energizing relay coil *1CR*. Relay contact $1CR_1$ closes, interlocking around the *START* push-button switch. Relay coil *1CR* now remains energized until such time as the temperature of the heated medium drops below 80° F. At that time, contactor coil *10CR* will again energize, energizing the heating elements and opening the *10CR* auxiliary contact. Relay coil *1CR* will be deenergized.

Time Control

Operation

Many types of timers are available for use on industrial machines. Their major function is to place information about elapsed time into an electric control circuit. The method of accomplishing this varies with the type of timer used. Time range, accuracy, and contact arrangement vary among types of timers. A few suggestions are made here for the proper use of timers. It is true that time elapses during nearly every action taking place on a machine. This does not necessarily mean that a timer is always the best means of control.

CONDUCTORS CARRYING LOAD CURRENT AT LINE VOLTAGE ARE DENOTED BY HEAVY LINES.

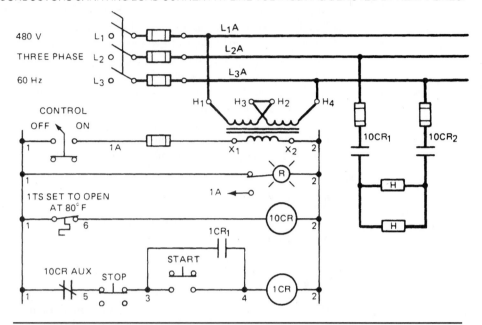

Figure 2–47   Control circuit.

For example, a machine part is to move from point X to point Y. This is a definite distance. It is observed that this motion requires approximately 5 seconds to complete. If the position of the machine part is important, a position control device (such as a limit switch) should be used to indicate that the part did arrive at point Y, not dependent on elapsed time. Due to variables in the machine, one cycle could require 5.0 seconds, the next 5.1 seconds, and a third cycle 4.9 seconds. Thus, if a timer is used, the machine part will not always stop at point Y.

A situation may exist in which a timer is substituted for a pressure indication. In one case, the pressure may build to a preset value before a preset time has elapsed. Also, time may elapse before the preset pressure is reached. A like example is when a timer is substituted for a temperature controller to obtain temperature information. The important point is to determine the critical condition to be met. Is it time, position, pressure, or temperature? Timers are very useful tools, but they must be applied properly.

Another point to clarify is reference to timers and time-delay relays. Generally, *time-delay relays* are devices having a timing function after the timer coil has been energized or deenergized. Time-delay relays have a normally open and a normally closed timing contact. In some cases, the contacts are isolated. In other cases, they have a common terminal. Sometimes time-delay relays have several normally open and/or normally closed instantaneous contacts that operate immediately when the time-delay coil is energized.

When reference is made to *timers*, the time function may start on one or more of the contacts upon energization or at any time after energization during a preset time cycle. Likewise, the timing function may stop on one contact or more during the cycle after timing has been started on the particular contact(s). In general, a timer opens or closes electric circuits to selected operations according to a timed program. Instantaneous contacts are also available on some timers.

Timing functions start from an electric signal, initiated through any one of several components. This may be a push-button switch, relay contact, temperature switch, pressure switch, limit switch, or so on. Once an electric signal is available to the timer, the timing function can in general be one of two types: *ON delay* (timing function after energizing) or *OFF delay* (timing function after deenergizing).

Other forms of this basic timing function are found with the solid-state timer, including *pulse* and *pulse-and-repeat* operations. In some timers, an *inhibit function* is available. This is accomplished by applying a voltage to specific terminals. The timing cycle then stops and holds its outputs in the last state without resetting the circuit. Timing continues and the outputs are allowed to change according to their programmed operating modes when voltage is removed from the aforementioned specific terminals.

Types of Timers

In grouping timers according to their method of timing, there are two types that are applied most frequently on industrial machines:

1. Synchronous motor-driven
2. Solid state

Four other types of timers that have industrial applications are the dashpot, mechanical, electrochemical, and thermal. Since these have only limited use on industrial machines, they are not covered in this text.

Synchronous Motor-Driven Timers   The *synchronous motor-driven timer* can well be termed the *workhorse* of the timer field. Its many applications make it a very useful tool. Under the general heading of synchronous motor-driven timers, the following subdivisions can be made:

- Reset timers
- Repeat-cycle timers
- Manual-set timers

The *reset timer* depends on the *clutch* and the *synchronous motor* for its operation. The symbols for the clutch and synchronous motor are shown in Figure 2–48. To provide an output for the timer to control electric circuits, most reset timers have a set of three contacts. Each contact can be adjusted to perform in the same way or in a different way, as required.

The circuit in Figure 2–49(A) illustrates conditions in which the motor is energized directly from the clutch being energized. This is the normal usage. Figure 2–49(B) illustrates the motor energized at a later time after the clutch is energized. Increased timing accuracy is possible with arrangement if the timing period is relatively short in an overall long cycle.

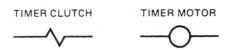

Figure 2–48   Symbols for timer clutch and timer motor.

Some of the simpler reset timers have only a snap-action switch that operates at the end of a preset time period. When the clutch is deenergized, the timer resets and the switch is released.

Another form of the reset timer is often called a *multiple-interval timer.* It is used extensively in programming control. Multiple time periods can be controlled by the adjustment of each individual *ON* and *OFF* setting.

The *manual-set timer* requires manual operation of the timer to start operation. The timer then runs a selected time and stops automatically.

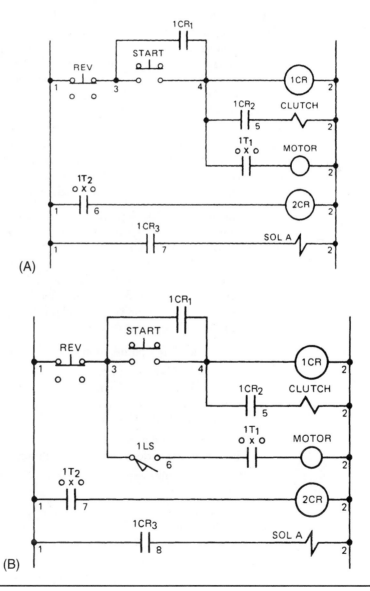

(A)

(B)

Figure 2–49 Control circuits.

Solid-State Timers Great advancement has been made in timer design in the last ten years. A thorough review of the many types and variations of *solid-state timers* would more than fill this entire book. For our purposes, then, we shall discuss the following important design features:

- Microprocessor-based timers
- Digital set and optional digital readout
- Provision for external set

In the *microprocessor-based timer/counter,* time or count operation, time range, and standard or reverse operation are selected via seven rocker switches located within the unit. Time and count functions are entered through a keypad. The digital display normally flashes while the unit is counting or timing.

Typical programmed outputs can be programmed to operate in one of four load sequences: oox, oxo, oox with pulse output, and oox pulse output with repeat-cycle operation.

The setting accuracy for count is 100 percent. The setting accuracy for time is 0.05 percent or 50 milliseconds, whichever is larger. The repeat accuracy for count is 100 percent. For time it is .001 percent or 35 milliseconds, whichever is larger.

The *SET* and *ENT* keys on the front panel provide access to the set point and to the front panel programmed functions. Programming changes are entered using the *increment* and *decrement* keys. A keypad *lock* function is built into the software of the unit, which allows the set point to be viewed but does not allow unauthorized changes. The timing cycle progress is shown on four 0.3-inch red LED displays for easy readability.

Circuit Applications Refer again to Figure 2–42(C). The ram advances until it reaches the work piece and builds pressure. On reaching a preset pressure, the normally closed pressure switch contact opens, and the ram reverses.

In Figure 2–50, a reset timer is added. The pressure switch contact is changed from normally closed to normally open. Note that the timer contacts are not controlled by a pneumatic or synchronous motor-driven timer, which is normally shown in a schematic. They are controlled from a microprocessor-based timer.

The sequence of operations for the circuit shown in Figure 2–50 is as follows:

1. Operate the *START* push-button switch.
2. Relay coil *1CR* is energized.
   (a) Relay contact $1CR_1$ closes, interlocking around *START* push-button switch.
   (b) Relay contact $1CR_2$ closes, energizing solenoid *A*. The piston advances to the work piece and builds pressure.
3. The normally open pressure switch contact closes, energizing the timer clutch. Timer contact $1T_1$ closes, energizing the timer motor.
4. When a preset time elapses, timer contact $1T_1$ opens, deenergizing the timer motor.
5. Timer contact $1T_2$ closes, energizing relay *2CR*.
   (a) Relay contact $2CR_1$ opens, deenergizing relay coil *1CR*.
   (b) Relay contact $1CR_1$ opens, opening the interlock circuit around the *START* push-button circuit.
   (c) Relay contact $1CR_2$ opens, deenergizing solenoid *A*.
6. Pressure drops, opening pressure switch contact *PS*.

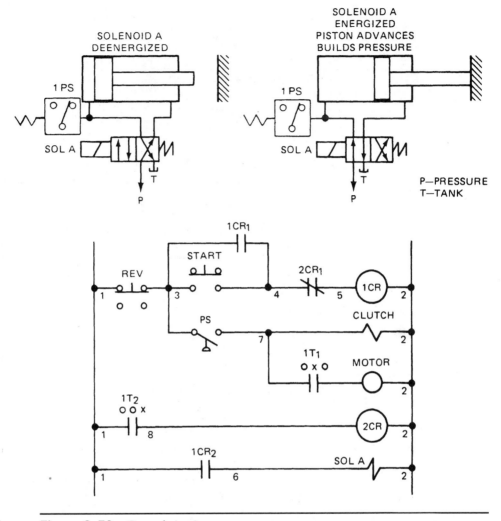

Figure 2–50 Control circuit.

7. Timer clutch deenergized.

8. Timer contact $1T_2$ opens, deenergizing relay *2CR*.

9. Normally closed relay contact $2CR_1$ closes, setting up the circuit for the next cycle.

Count Control Preset Electrical Impulses

The electromechanical control counter is similar to the electromechanical reset timer, except that the synchronous motor of the timer is replaced by a stepping motor. The stepping motor advances one step each time it is deenergized. After a preset number of electric impulses, a contact is opened or closed. The electromechanical counter requires a minimum of approximately 0.5 seconds OFF time between input impulses to reset. As in the electromechanical timer, there are three output contacts. Each contact can be arranged for a sequence.

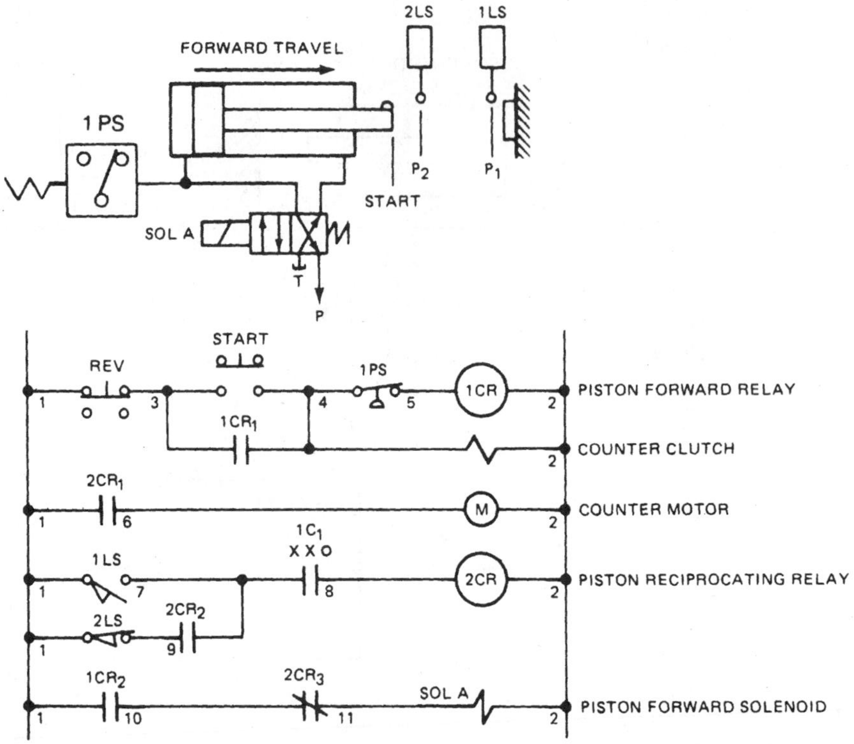

Figure 2–51 Control circuit.

With the clutch deenergized, the contacts are in the *reset condition*. When the clutch is energized, the contacts go to the *counting condition*. When the count motor has received the same number of count impulses as set on the dial, the contacts go to the *counted-out condition*. When the clutch is deenergized, the contacts return to the reset on.

Timing devices are generally available with analog set and readout dials.

Circuit Applications

A typical control circuit using a reset counter is shown in Figure 2–51. In this example a single piston-cylinder assembly is used. The piston is to travel to the right as shown until it engages and operates limit switch *1LS* at position P_1.

The piston returns to position P_2, operating limit switch *2LS*. The piston now travels to the right. This reciprocating motion continues until the preset number of counts has been reached. The counter is now in the counted-out condition. The piston continues to the right until a work piece is engaged and preset pressure builds to the setting of pressure switch *1PS*. The piston now returns to the start position.

A single-solenoid, spring-return operating valve is used to supply fluid power to the cylinder. The circuit is arranged to return the piston to the start position at any time by operating the *REVERSE* push-button switch.

The sequence of operation proceeds as follows:

1. Operate the *START* push-button switch.
2. Relay coil *1CR* is energized.

(a) Relay contact $1CR_1$ closes, interlocking around the *START* push-button switch and energizing the counter clutch.

(b) Relay contact $1CR_2$ closes, energizing solenoid *A*.

The piston travels to the right. At position P_2 limit switch $2LS$ is operated. No action results as relay contact $2CR_2$ is open. The piston continues to travel to position P_1, operating limit switch $1LS$.

3. Relay coil $2CR$ is energized.

(a) Relay contact $2CR_1$ closes, energizing the counter motor.

(b) Relay contact $2CR_2$ closes, forming an interlock circuit with normally closed limit switch contact $2LS$ around normally open limit switch contact $1LS$.

(c) Relay contact $2CR_3$ opens, deenergizing solenoid *A*. The piston travels back (to the left) until it operates limit switch $2LS$.

4. Relay coil $2CR$ is deenergized.

(a) Relay contact $2CR_1$ opens, deenergizing the counter motor. The counter has now completed one count.

(b) Relay contact $2CR_2$ opens.

(c) Relay contact $2CR_3$ closes, energizing solenoid *A*.

The piston now travels forward (to the right) until it operates limit switch $1TS$.

5. Relay coil $2CR$ energized.

(a) Relay contact $2CR_1$ closes, energizing the counter motor.

(b) Relay contact $2CR_2$ closes, interlocking limit switch $1LS$.

(c) Relay contact $2CR_3$ opens, deenergizing solenoid *A*.

The piston travels back until it operates limit switch $2LS$.

6. Relay coil $2CR$ is deenergized.

(a) Relay contact $2CR_1$ opens, deenergizing the counter motor. The counter has now completed two counts.

(b) Relay contact $2CR_2$ opens.

(c) Relay contact $2CR_3$ closes, energizing solenoid *A*.

The piston continues to shuttle between limit switches $1LS$ and $2LS$ until the counter has counted the number of preset counts. When the counter has counted out, counter contact $1C_1$ opens. The next time the piston advances and operates limit switch $1LS$, relay coil $2CR$ is not energized.

7. Solenoid *A* remains energized.

8. Piston continues past limit switch $1LS$, meeting the work piece and introduces pressure to a preset amount on pressure switch $1PS$.

9. The normally closed pressure switch contact $1PS$ opens.

10. Relay coil $1CR$ is deenergized.

(a) Relay contact $1CR_1$ opens, opening the interlock circuit around the *START* push-button switch and deenergizing the counter clutch.

(b) Contact $1CR_2$ opens, deenergizing solenoid *A*.

The piston now returns to its initial start position.

Solid-State Counters

Considerable advancement has been made in solid-state counter design. *Solid-state counters* are available with *digital set* and *digital readout*. High-speed pulse operation with 100 percent accuracy is available.

Three counting modes can be programmed: count directional, add/subtract, and quadrature. This unit is able to count up from zero or count down from the set point. The counter can be programmed to give a pulse at count-out and automatically reset, or it can be programmed for single-cycle latch operation. The counter is programmed with six dip switches located on the side of the counter.

Counter-output action occurs when the count total indicated by front-mounted thumbwheel switches is reached. The counter sets to the selected thumbwheel setting when power is applied to terminals A and B. Counts are applied to a count-input terminal, and each count is registered on contact opening. When registered counts equal the set point, the output changes state. The output remains in this state as long as the line voltage is applied to terminals A and B.

A *count inhibit* is available with this counter. When line voltage is applied to the count inhibit from either terminals A or B, incoming count pulses are not counted. The counter remembers count total at the time that the inhibit is applied and resumes counting from that point after the inhibit voltage is removed. The count inhibits can be applied from either side of the powerline.

CHAPTER

3

Mechanical Motor Control

Magnetic Clutches

Electrically Controlled Magnetic Clutches

Machinery clutches were originally designed to engage very large motors to their loads after the motors had reached running speeds. *Electronically controlled magnetic clutches* provide smooth starts for operations in which the material being processed might be damaged by abrupt starts (Figure 3–1). Clutches are also used to start high-inertia loads, since the starting may be difficult for a motor that is sized to handle the running load. When starting conditions are severe, a clutch inserted between the motor and the load allows the motor to run within its load capacity. The motor will take longer to bring the load up to speed, but the motor and load will not be damaged. As more automatic cycling and faster cycling rates are being required in industrial production, electrically controlled clutches are being used more often.

Single-Face Clutch: The *single-face clutch* consists of two disks: One is the field member (electromagnet), and the other is the armature member. The operation of the clutch is similar to that of the electromagnet in a motor starter (Figure 3–1). When current is applied to the field-winding disk through collector (slip) rings, the two disks are drawn together magnetically. The friction face of the field disk is held tightly against the armature disk to provide positive engagement between the rotating drives. When the current is removed, a spring action separates the faces to provide a definite clearance between the disks. In this manner, the motor is mechanically disconnected from the load.

Multiple-Face Clutch *Multiple-face clutches* are also available. In a double-face clutch, both the armature and the field disks are mounted on a single hub with a double-faced friction lining supported between them. When the magnet of the field member is energized, the armature and field members are drawn together. They grip the lining between them to provide the driving torque. When the magnet is de energized, a spring separates the two members and they rotate independently of each other. Double-face clutches are available in sizes up to 78 inches in diameter.

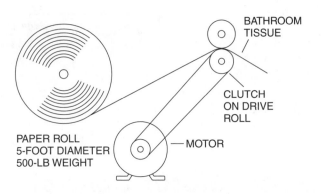

Figure 3–1   To prevent tearing, a cushioned start is required on a drive roll that winds bathroom tissue off a large roll. The roll is 5 feet in diameter and weighs 500 pounds when full. Pick-up of thin tissue must be very gradual to avoid tearing. Application also can be used for filmstrip-processing machine.

A *water-cooled magnetic clutch* is available for applications that require a high degree of slippage between the input and output rotating members. Uses for this type of clutch include tension control (windup and payoff) and cycling (starting and stopping) operations in which large differences between the input and the output speeds are required. Flowing water removes the heat generated by the continued slippage within the clutch. A rotary water union mounted in the end of the rotor shaft means that the water-cooled clutch cannot be end-coupled directly to the prime mover. Chains or gears must be used.

A combination clutch and magnetic brake disconnects the load from the drive and simultaneously applies a brake to the load side of the drive. Magnetic clutches and brakes are often used as mechanical power-switching devices in module form. Remember that the quicker the start or stop, the shorter the life of this equipment.

Magnetic clutches are used on automatic machines for starting, running, cycling, and torque limiting. The combinations and variations of these functions are practically limitless.

Magnetic Drives

The *magnetic drive* couples the motor to the load magnetically. The magnetic drive can be used as a clutch and can be adapted to an adjustable-speed drive. *Electromagnetic (or eddy current) coupling* is one of the simpler ways to obtain an adjustable output speed from the constant input speed of squirrel-cage motors.

There is no mechanical contact between the rotating members of the magnetic drive. Thus, there is no wear. Torque is transmitted between the two rotating units by an electromagnetic reaction created by an energized coil winding. The slip between the motor and load can be controlled continuously with more precision and over a wider range than is possible with the mechanical friction clutch.

As shown in Figure 3–2, the magnet rotates within the steel ring or drum. There is an air gap between the ring and the magnet. The magnetic flux crosses the air gap and penetrates the iron ring. The rotation of the ring with relation to the magnet generates eddy currents and magnetic fields in the ring. Magnetic interaction between the two units transmits torque from the motor to the load. This torque is controlled with a rheostat that manually or automatically adjusts the DC supplied to the electromagnet through the slip rings.

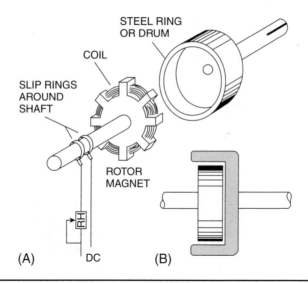

Figure 3–2 Diagram showing (A) open view of magnetic drive assembly. (B) Spider rotor magnet rotates within ring.

When the electromagnetic drive responds to an input or command voltage, a further refinement can be obtained in automatic control to regulate and maintain the output speed. The magnetic drive can be used with any type of actuating device or transducer that can provide an electric signal. For example, electronic controls and sensors that detect liquid level, air and fluid pressure, temperature, and frequency can provide the input required.

A *tachometer generator* provides feedback speed control in that it generates a voltage that is proportional to its speed. Any changes in load condition will change the speed. The resulting generator voltage fluctuations are fed to a control circuit that increases or decreases the magnetic-drive-field excitation to hold the speed constant.

For applications where a magnetic drive meets the requirements, an adjustable speed is frequently a desirable choice. Magnetic drives are used for applications requiring an adjustable speed, such as cranes, hoists, fans, compressors, and pumps.

Direct Drives

Directly Coupled Drive Installation

The most economical speed for an electric motor is about 1800 rpm. Most electrically driven-constant speed machines, however, operate at speeds below 1800 rpm. These machines must be provided with either a high-speed motor and some form of mechanical speed reducer or a low-speed, directly coupled motor.

Synchronous motors can be adapted for direct coupling to machines operating at speeds from 3600 rpm to about 80 rpm, with horsepower ratings ranging from 20 to 5000 and above. It has been suggested that synchronous motors are less expensive to install than squirrel-cage motors if the rating exceeds one horsepower. However, this recommendation considers only the first cost. It does not take into account the higher efficiency and better power factor of the synchronous motor. When the motor speed matches the machine input shaft speed, a simple mechanical coupling is used, preferably a flexible coupling.

Trouble-free operation can usually be obtained by following several basic recommendations for the installation of directly coupled drives and pulley or chain drives. First, the motor and machine must be installed in a level position. When connecting the motor to its load, the alignment of the devices must be checked more than once from positions at right angles to each other. For example, when viewed from the side, two shafts may appear to be in line. When the same shafts are viewed from the top, as shown in Figure 3–3A, it is evident that the motor shaft is at an angle to the other shaft. A dial indicator should be used to check the alignment of the motor and the driven machinery Figure 3–3B. If a dial indicator is not available, a feeler gauge may be used.

During the installation, the shafts of the motor and the driven machine must be checked to ensure that they are not bent. Both the machine and the motor should be rotated together, just as they rotate when the machine is running, and then rechecked for alignment. After the angle of the shafts is aligned, the shafts may appear to share the same axis. However, as shown in Figure 3–3B, the axes of the motor and the driven machine may really be off center. When viewed again from a position 90 degrees away from the original position, it can be seen that the shafts are not on the same axis.

To complete the alignment of the devices, the motor should be moved until rotation of both shafts shows that they share the same axis when viewed from four positions spaced 90 degrees apart around the shafts. The final test is to check the starting and running currents with the connected load to ensure that they do not exceed specifications. There are several disadvantages to the use of low-speed, directly coupled induction motors. They usually have a low power factor and low efficiency. Both of these characteristics increase electric power costs. Because of this, induction motors are rarely used for operation at speeds below 500 rpm.

MOTOR   LOAD

Figure 3–3A   Every alignment check must be made from positions 90 degrees apart or at right angles to each other.

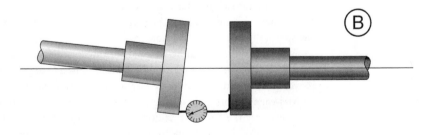

Figure 3–3B   Angular check of direct motor couplings.

Constant-speed motors are available with a variety of speed ratings. The highest possible speed is generally selected to reduce the size, weight, and cost of the motor. At 5 horsepower, a 1200-rpm motor is almost 50 percent larger than an 1800-rpm motor. At 600 rpm, the motor is well over twice as large as the 1800-rpm motor. In the range from 1200 rpm to 900 rpm, the size and cost disadvantages may not be overwhelming factors. Where this is true, low-speed, directly coupled motors can be used. For example, this type of motor is used on most fans, pumps, and compressors.

Pulley Drives

Installation

Flat belts, V-belts, chains, or gears are used on motors so that smooth speed changes at a constant rpm can be achieved. For speeds below 900 rpm, it is practical to use an 1800-rpm or 1200-rpm motor connected to the driven machine by a V-belt or a flat belt. Machine shafts and bearings give long service when the power-transmission devices are properly installed according to the manufacturer's instructions.

Offset drives, such as V-belts, gears, and chain drives, can be lined up more easily than direct drives. Both the motor and the load shafts must be level. A straightedge can be used to ensure that the motor is aligned on its axis and that it is at the proper angular position so that the pulley sheaves of the motor and the load are in line. When belts are installed, they should be tightened just enough to ensure nonslip engagement. The less cross tension there is, the less wear there will be on the bearings involved. Proper and firm positioning and alignment are necessary to control the forces that cause vibration and the forces that cause thrust.

The designer of a driven machine usually determines the motor mount and the type of drive to be used. This means that the installer has little choice in the motor location. In many flat-belt or V-belt applications, however, the construction or maintenance electrician may be called upon to make several choices. If a choice can be made, the motor should be placed where the force of gravity helps to increase the grip of the belts. A *vertical drive* can cause problems because gravity tends to pull the belts away from the lower sheave. To counteract this action, the belts require far more tension than the bearings should have to withstand. The electrician should avoid this type of installation, if possible.

There are correct and incorrect placements for *horizontal drives.* The location where the motor is to be installed can be determined from the direction of rotation of the motor shaft. It is recommended that the motor be placed *with* the direction of rotation so that the belt slack is on top. In this position, the belt tends to wrap around the sheaves. This problem is less acute with V-belts than it is with flat belts or chains. Therefore, if rotation is to take place in both directions, V-belts should be used. In addition, the motor should be placed on the side of the most frequent direction of rotation.

Pulley Speeds

Motors and machines are frequently shipped without pulleys or with pulleys of incorrect sizes. The drive and driven speeds are given either on the motor and machine nameplates or in the descriptive literature accompanying the machines.

Four quantities must be known if the machinery is to be set up with the correct pulley sizes: the drive revolutions per minute, the driven revolutions per minute, the diameter of the drive pulley, and the diameter of the driven pulley. If three of these quantities are known, the fourth quantity can be determined. For example, if a motor runs at 3600 rpm, the driven speed is 400 rpm, and there is a 4-inch pulley for the motor, the size of the pulley for the driven load can be determined from the following equation:

$$\frac{Drive\ rpm}{Driven\ rpm} = \frac{driven\ pulley\ diameter}{drive\ pulley\ diameter}$$

$$\frac{3600}{400} = \frac{x}{4}$$

By cross multiplying and then dividing, we arrive at the pulley size required:

$$4x = 144$$

$$x = \frac{144}{4} = 36\text{-}inch\text{-}diameter\ pulley$$

If both the drive and the driven pulleys are missing, the problem can be solved by estimating a reasonable pulley diameter for either pulley and then using the equation with this value to find the fourth quantity.

CHAPTER

4

Digital-Logic Motor Control

Digital Logic   The electrician in today's industry must be familiar with solid-state, *digital-logic* circuits. *Digital* refers to a device that has only two states, on or off. Most electricians have been using digital logic for many years without realizing it. Magnetic relays, for instance, are digital devices. Relays are generally considered to be single-input, multi-output devices. The coil is the input, and the contacts are the output. A relay has only one coil, but it may have a large number of contacts.

Although relays are digital devices, the term *digital logic* refers to circuits that use solid-state control devices known as *gates*. There are five basic types of gates: the AND, OR, NOR, NAND, and INVERTER. Each of these gates is covered later in this chapter.

There are also different types of logic. For instance, one of the earliest types of logic to appear was *resistor-transistor logic (RTL)*. This was followed by *diode-transistor logic (DTL)*, and *transistor-transistor logic (TTL)*. Resistor-transistor logic and diode-transistor logic are not used much anymore, but transistor-transistor logic is still used to a fairly large extent. The latter can be identified because it operates on 5 volts.

One type of logic frequently used in industry is *high-transit logic (HTL)*. High-transit logic is used because it does a better job of ignoring the voltage spikes and drops caused by the starting and stopping of inductive devices such as motors. High-transit logic generally operates on 15 volts.

Another type of logic that has very high input impedance and has become very popular is known as *CMOS*, which comes from *complementary-symmetry metal-oxide-semiconductor (COSMOS)*. The advantage of CMOS logic is that it requires very little power to operate; however, there are also some disadvantages. One disadvantage is that CMOS logic is so sensitive to voltage that the static charge of a person's body can sometimes destroy an integrated circuit just by touching it. People that work with CMOS logic often use a ground strap that straps around the wrist like a bracelet. This strap is used to prevent a static charge from building up on the body. Another characteristic of CMOS logic is that unused inputs cannot be left in an indeterminate state. Unused inputs must be connected to either a high state or a low state.

The AND Gate

While magnetic relays are single-input, multi-output devices, gate circuits are multi-input, single-output devices. For instance, an *AND gate* may have several inputs but only one output. Figure 4–1A shows the USA Standards Institute (USASI) symbol for an AND gate with three inputs labeled *A, B,* and *C* and one output labeled *Y.*

USASI symbols are more commonly referred to as *computer-logic symbols.* Unfortunately for industrial electricians, there is another system known as *NEMA logic* that uses a completely different set of symbols. The NEMA symbol for a three-input AND gate is shown in Figure 4–1B.

Although both symbols mean the same thing, they are drawn differently. Electricians working in industry must learn both sets of symbols because both types of symbols are used. Regardless of which type of symbol is used, the AND gate operates the same way. An AND gate must have all of its inputs high in order to get an output. If it is assumed that transistor-transistor logic is being used, a high level is considered to be +5 volts and a low level is considered to be 0 volts. Figure 4–1C shows the truth table for a two-input AND gate.

The *truth table* is used to illustrate the state of a gate's output with different conditions of input. The number *one* represents a high state, and *zero* represents a low state. Notice in Figure 4–1C that the output of the AND gate is high only when both of its inputs are high. The operation of the AND gate is very similar to that of the simple relay circuit shown in Figure 4–1D.

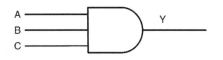

Figure 4–1A USASI symbol for a three-input AND gate.

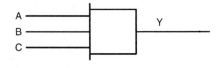

Figure 4–1B NEMA logic symbol for a three-input AND gate.

A	B	Y
0	0	0
0	1	0
1	0	0
1	1	1

Figure 4–1C Truth table for a two-input AND gate.

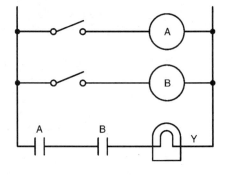

Figure 4–1D Equivalent relay circuit for a two-input AND gate.

If a lamp is used to indicate the output of the AND gate, both relay coils *A* and *B* must be energized before there can be an output. Figure 4–1E shows the truth table for a three-input AND gate. Notice that there is still only one condition that permits a high output for the gate, and that condition is when all inputs are high or at logic level *one*. *When using an AND gate, any zero input = a zero output.* An equivalent relay circuit for a three-input AND gate is shown in Figure 4–1F.

The OR Gate

The computer-logic symbol and the NEMA-logic symbol for the *OR gate* are shown in Figure 4–2A. The OR gate has a high output when either or both of its inputs are high. Refer to the truth table shown in Figure 4–2B. An easy way to remember how an OR gate functions is to say that *any* one *input* = *a* one *output*. An equivalent relay circuit for the OR gate is shown in figure 4–2C. Notice in this circuit that if either or both of the relays are energized, there will be an output at *Y.*

The EXCLUSIVE-OR Gate   Another gate that is very similar to the OR gate is known as an *EXCLUSIVE-OR gate.* The symbol for an EXCLUSIVE-OR gate is shown in Figure 4–3A. The EXCLUSIVE-OR gate has a high output when either, but not both, of its inputs are high. Refer to the truth table shown in Figure 4–3B. An equivalent relay circuit for the EXCLUSIVE-OR gate is shown in Figure 4–3C. Notice that if both relays are energized or deenergized at the same time, there is no output.

The INVERTER

The simplest of all the gates is the *INVERTER.* The INVERTER has one input and one output. As its name implies, the output is *inverted,* or the opposite of the input. For example,

A	B	C	Y
0	0	0	0
0	0	1	0
0	1	0	0
0	1	1	0
1	0	0	0
1	0	1	0
1	1	0	0
1	1	1	1

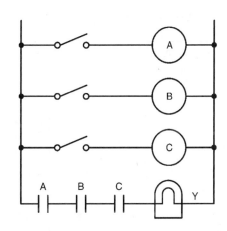

Figure 4–1E   Truth table for a three-input AND gate.

Figure 4–1F   Equivalent relay circuit for a two-input AND gate.

Figure 4–2A Computer-logic symbol and NEMA-logic symbol for an OR gate.

A	B	Y
0	0	0
0	1	1
1	0	1
1	1	1

Figure 4–2B Truth table for an OR gate.

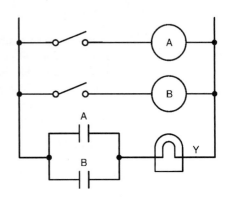

Figure 4–2C Equivalent relay circuit for an OR gate.

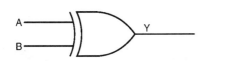

Figure 4–3A EXCLUSIVE-OR gate computer-logic symbol.

A	B	Y
0	0	0
0	1	1
1	0	1
1	1	0

Figure 4–3B Truth table for an EXCLUSIVE-OR gate.

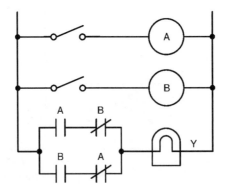

Figure 4–3C Equivalent relay circuit for an EXCLUSIVE-OR gate.

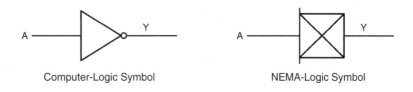

Figure 4–4A Computer-logic symbol and NEMA-logic symbol for an INVERTER.

A	Y
0	1
1	0

Figure 4–4B Truth table for an INVERTER.

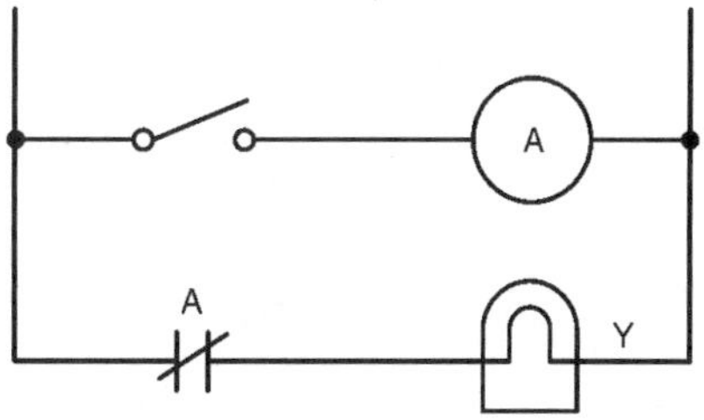

Figure 4–4C Equivalent relay circuit for an INVERTER.

if the input is high, the output is low, or if the input is low, the output is high. Figure 4–4A shows the computer-logic and NEMA symbols for an INVERTER.

In computer logic, a circle drawn on a gate means to invert. Since the *0* appears on the output end of the gate, it means the output is inverted. In NEMA logic, an *X* is used to show that a gate is inverted. The truth table for an INVERTER is shown in Figure 4–4B. The truth table clearly shows that the output of the INVERTER is the opposite of the input. Figure 4–4C shows an equivalent relay circuit for the INVERTER.

The NOR Gate

The *NOR* gate is the *NOT OR* gate. Referring to the computer-logic and NEMA-logic symbols for a NOR gate in Figure 4–5A, notice that the symbol for the NOR gate is the same as the symbol for the OR gate with an inverted output. A NOR gate can be made by connecting an INVERTER to the output of an OR gate as shown in Figure 4–5B.

The truth table shown in Figure 4–5C shows that the output of a NOR gate is *zero,* or low, when any input is high. Therefore, it could be said that *any* one *input = a* zero *output* for the NOR gate. An equivalent relay circuit for the NOR gate is shown in Figure 4–5D. Notice in Figure 4–5D that if either relay *A* or *B* is energized, there is no output at *Y.*

The NAND Gate

The *NAND gate* is the *NOT AND* gate. Figure 4–6A shows the computer-logic symbol and the NEMA-logic symbol for the NAND gate. Notice that these symbols are the same as the symbols for the AND gate with inverted outputs. If any input of a NAND gate is low, the output is high. Refer to the truth table in Figure 4–6B. Notice that the truth table clearly indicates that *any* zero *input = a* one *output.* Figure 4–6C shows an equivalent relay circuit for the NAND gate. If either relay *A* or relay *B* is deenergized, there is an output at *Y.*

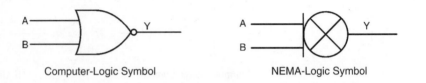

Computer-Logic Symbol   NEMA-Logic Symbol

Figure 4–5A Computer-logic symbol and NEMA-logic symbol for a two-input NOR gate.

Figure 4–5B Equivalent relay circuit for a NOR gate.

A	B	Y
0	0	1
0	1	0
1	0	0
1	1	0

Figure 4–5C Truth table for a two-input NOR gate.

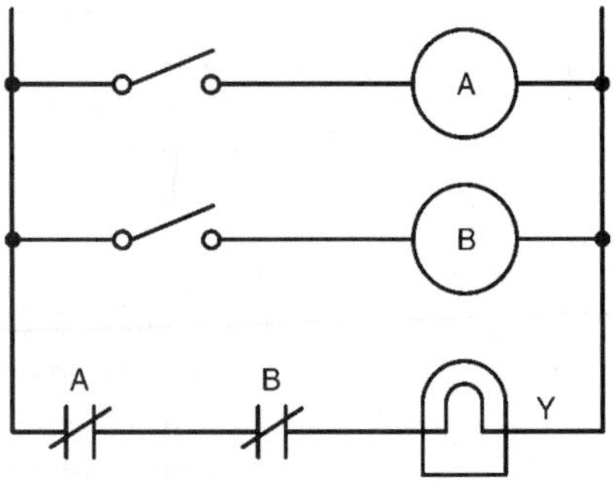

Figure 4–5D Equivalent relay circuit for a two-input NOR gate.

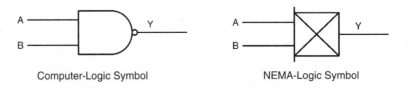

Computer-Logic Symbol   NEMA-Logic Symbol

Figure 4–6A Computer-logic symbol and NEMA-logic symbol for a two-input NAND gate.

The NAND gate is often referred to as the *basic gate* because it can be used to make any of the other gates. For instance, Figure 4–6D shows the NAND gate connected to make an INVERTER. If a NAND gate is used as an INVERTER and is connected to the output of another NAND gate, it will become an AND gate as shown in Figure 4–6E. When two NAND gates are connected as INVERTERS and these INVERTERS are connected to the inputs of another NAND gate, an OR gate is formed, Figure 4–6F. If an INVERTER is added to the output of the OR gate shown in Figure 4–6F, a NOR gate is formed (Figure 4–6G).

A	B	Y
0	0	1
0	1	1
1	0	1
1	1	0

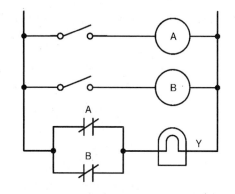

Figure 4–6B Truth table for a two-input NAND gate.

Figure 4–6C Equivalent relay circuit for a two-input NAND gate.

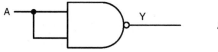

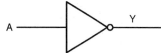

Figure 4–6D NAND gates connected as an INVERTER.

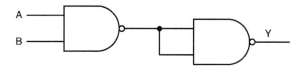

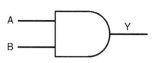

Figure 4–6E NAND gates connected as an AND gate.

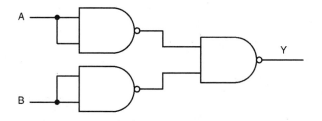

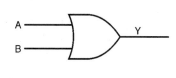

Figure 4–6F NAND gates connected as an OR gate.

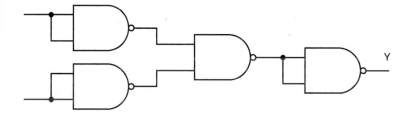

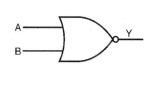

Figure 4–6G NAND gates connected as a NOR gate.

Integrated Circuits

Digital-logic gates are generally housed in fourteen-pin, *integrated-circuit* packages. One of the old reliable types of transistor-transistor logic that is frequently used is the *7400 family of devices.* For instance, a 7400 integrated circuit is a quad, two-input, positive NAND gate. The word *quad* means that there are four NAND gates contained in the package. Each NAND gate has *two inputs,* and *positive* means that a level *one* is considered to be a positive voltage. There can, however, be a difference in the way integrated circuits are connected. A 7400 (J or N) integrated circuit has a different pin connection than a 7400 (W) package.

Testing Integrated Circuits

Integrated circuits cannot be tested with a volt-ohm-milliammeter. Most integrated circuits must be tested by connecting power to them and then testing the inputs and outputs with special test equipment. Most industrial equipment is designed with different sections of the control system built in modular form. The electrician determines which section of the circuit is not operating and replaces that module. The defective module is then sent to the electronics department or to a company outside of the plant for repair.

The Bounceless Switch

When a control circuit is constructed, it must have *sensing devices* to tell it what to do. The number and type of sensing devices used are determined by the circuit. Sensing devices can range from a simple push button to float switches, limit switches, and pressure switches. Most of these sensing devices use some type of mechanical switch to indicate their condition. A *float switch,* for example, indicates its condition by opening or closing a set of contacts. The float switch can *tell* the control circuit that a liquid is either at a certain level or not. Most of the other types of sensing devices use this same method to indicate some condition. A *pressure switch* indicates that a pressure is either at a certain level or not, and a *limit switch* indicates if some device has moved a certain distance or if a device is present or absent from some location.

Almost all of these devices employ a *snap-action switch.* When a mechanical switch is used, the snap action is generally obtained by spring-loading the contacts. This snap action is necessary to ensure good contact when the switch operates. Assume that a float switch is used to sense when water reaches a certain level in a tank. If the water rises at a slow rate, the contacts will come together at a slow rate, resulting in a poor connection. However, if the contacts are spring-loaded, when the water reaches a certain level, the contacts will snap from one position to another.

Although most contacts have a snap action, they do not generally close with a single action. When the movable contacts meet the stationary contact, there is often a fast *bouncing* action. This means that the contacts may actually make and break contact three or four times in succession before the switch remains closed. When this type of switch is used to control a relay, contact bounce does not cause a problem because relays are relatively slow-acting devices (Figure 4–7).

Figure 4–7 Contact bounce switch does not greatly affect relay circuits.

When this type of switch is used with an electronic control system, however, contact bounce can cause a great deal of trouble. Most digital-logic circuits are very fast-acting and can count each pulse when a contact bounces. Depending on the specific circuit, each of these pulses may be interpreted as a command. Contact bounce can cause the control circuit to literally *lose its mind.*

Since contact bounce can cause trouble in an electronic control circuit, contacts are debounced before they are permitted to *talk* to the control system. When contacts must be debounced, a circuit called a *bounceless switch* is used. Several circuits can be used to construct a bounceless switch, but the most common construction method uses digital-logic gates. Although any of the INVERTER gates can be used to construct a bounceless switch, in this example only two will be used.

Before construction of the circuit begins, the operation of a bounceless-switch circuit should first be discussed. The idea is to construct a circuit that will lock its output either high or low when it detects the first pulse from the mechanical switch. If its output is locked in a position, it will ignore any other pulse it receives from the switch. The output of the bounceless switch is connected to the input of the digital control circuit. The control circuit will now receive only one pulse instead of a series of pulses.

The first gate used to construct a bounceless switch is the INVERTER. The computer symbol and the truth table for the INVERTER are shown in Figure 4–8A. The bounceless switch circuit using INVERTERS is shown in Figure 4–8B. The output of the circuit should be high with the switch in the position shown. The switch connects the input of INVERTER *#1* directly to ground, or low. This causes the output of INVERTER *#1* to be at a high state. The output of INVERTER *#1* is connected to the input of INVERTER *#2*. Since the input of INVERTER *#2* is high, its output is low. The output of INVERTER *#2* is connected to the input of INVERTER *#1*. This causes a low condition to be maintained at the input of INVERTER *#1*.

A	Y
0	1
1	0

INVERTER Symbol   INVERTER Truth Table

Figure 4–8A Symbol and truth table for an INVERTER.

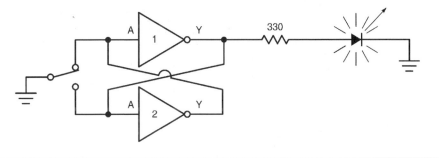

Figure 4–8B High output condition.

If the position of the switch is changed as shown in Figure 4–9, the output will change to low. The switch now connects the input of INVERTER #2 to ground, or low. The output of INVERTER #2 is, therefore, high. The high output of INVERTER #2 is connected to the input of INVERTER #1. Since the input connected to INVERTER #1 is now high, its output becomes low. The output of INVERTER #1 is connected to the input of INVERTER #2. This forces a low input to be maintained at INVERTER #2. Notice that the output of one INVERTER is used to lock the input of the other INVERTER.

The second logic gate used to construct a bounceless switch is the NAND gate. The computer symbol and the truth table for the NAND gate are shown in Figure 4–10A. The circuit in Figure 4–10B shows the construction of a bounceless switch using NAND gates. In this circuit, the switch has input *A* of gate #1 connected to low, or ground. Since input *A* is low, the output is high. The output of gate #1 is connected to input *A* of gate #2. Input *B* of gate #2 is connected to a high through the 4.7-kilohm resistor. Since both inputs of gate #2 are high, its output is low. This low output is connected to input *B* of gate #1. Since gate #1 now has a low connected to input *B,* its output is forced to remain high even if contact bounce causes a momentary high at input *A.*

When the switch changes position as shown in Figure 4–10C, input *B* of gate #2 is connected to a low. This forces the output of gate #2 to become high. The high output of gate #2 is connected to input *B* of gate #1. Input *A* of gate #1 is connected to a high through a 4.7-kilohmn resistor. Since both inputs of gate #1 are high, its output is low. This low is connected to input *A* of gate #2, which forces its output to remain high even if contact bounce causes a high to be momentarily connected to input *B.* The output of this circuit will remain constant even if the switch contacts bounce. The switch has now been debounced and is ready to be connected to the input of an electronic control circuit.

Start/Stop Push-Button Control*

In this section, a digital circuit will be designed to perform the same function as a common relay circuit. The relay circuit is a basic stop-start, push-button circuit with overload protection.

Before beginning the design of an electronic circuit that will perform the same function as this relay circuit, the operation of the relay circuit should first be discussed. In the circuit shown in Figure 4–11, no current can flow to relay coil *M* because the normally open start button and the normally open contact are controlled by relay coil *M.*

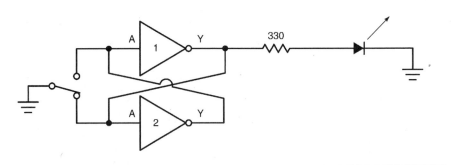

Figure 4–9   Low output condition.

*Adapted with permission from Herman/Alerich, *Industrial Motor Controls,* Delmar Publishers.

A	B	Y
0	0	1
0	1	1
1	0	1
1	1	0

NAND Symbol   NAND Truth Table

Figure 4–10A   Symbol and truth table for a NAND gate.

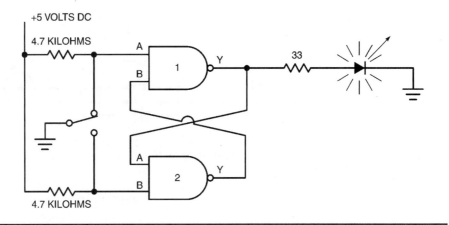

Figure 4–10B   High output condition.

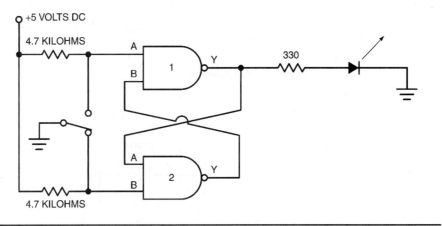

Figure 4–10C   Low output condition.

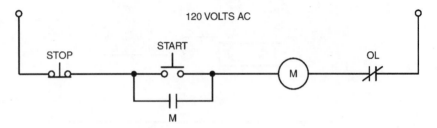

Start-stop, push-button circuit.

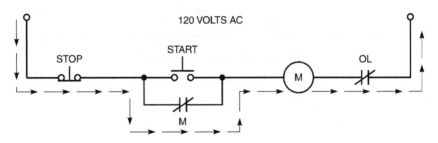

Start button energizes relay coil M.

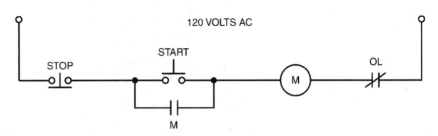

M contacts maintain the circuit.

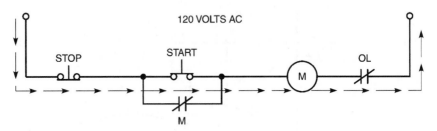

Stop button breaks the circuit.

Figure 4–11 Sequence of operation of start/stop station.

When the start button is pushed, current flows through the relay coil and normally closed overload contact to the power source. When current flows through relay coil M, the contacts connected parallel to the start button close. These contacts maintain the circuit to coil M when the start button releases and returns to its open position.

The circuit will continue to operate until the stop button is pushed and breaks the circuit to the coil. When the current flow to the coil stops, the relay deenergizes and contact M reopens. Since the start button is now open and contact M is open, there is no complete circuit to the relay coil when the stop button is returned to its normally closed position. If the relay is to be restarted, the start button must be pushed again to provide a complete circuit to the relay coil.

The only other logic condition that can occur in this circuit is caused by the motor connected to the load contacts of relay M. The motor is connected in series with the heater of an overload relay. When coil M energizes, it closes the load contact M as shown in the figure. When the load contact closes, it connects the motor to the 120-volt, AC powerline.

If the motor is overloaded, it will cause too much current to flow through the circuit. When a current greater than normal flows through the overload heater, the heater produces more heat than it does under normal conditions. If the current becomes high enough, it will cause the normally closed overload contact to open. Notice that the overload contact is electrically isolated from the heater. The contact, therefore, can be connected to a different voltage source than the motor.

If the overload contact opens, the control circuit is broken and the relay deenergizes as if the stop button had been pushed. After the overload contact has been reset to its normally closed position, the coil will remain deenergized until the start button is again pressed.

Now that the logic of the circuit is understood, a digital-logic circuit that will operate in this manner can be designed. The first problem is to find a circuit that can be turned on with one push button and turned off with another. The circuit shown in Figure 4–12A can perform this function. This circuit consists of an OR gate and an AND gate. Input A of the OR gate is connected to a normally open push button, which is connected to +5 volts DC. Input B of the OR gate is connected to the output of the AND gate. The output of the OR gate is connected to input A of the AND gate. Input B of the AND gate is connected through a normally closed push button to +5 volts DC. This normally closed push button is used as the stop button. The output of the AND gate is the output of the circuit.

To understand the logic of this circuit, assume that the output of the AND gate is low. This produces a low at input B of the OR gate. Since the push button connected to input A is open, a low is also produced at this input. When all inputs of an OR gate are low, its output is also low. The low output of the OR gate is connected to input A of the AND gate. Input B of the AND gate is connected to a high through the normally closed push-button switch. Since input A of the AND gate is low, the output of the AND gate is forced to remain in a low state.

When the start button is pushed, a high is connected to input A of the OR gate. This causes the output of the OR gate to change to high. This high output is connected to input A of the AND gate. The AND gate now has both of its inputs high, so its output changes from a low to a high state. When the output of the AND gate changes to a high state, input B of the OR gate also becomes high. Since the OR gate now has a high connected to its B input, its output will remain high when the push button is returned to its open condition and input A becomes low. Notice that this circuit operates the same as the relay circuit when

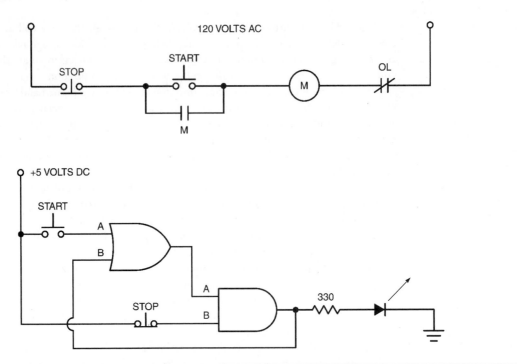

Figure 4–12A   Digital-logic circuit.

the start button is pushed. The output changes from a low state to a high state, and the circuit locks in this condition so the start button can be reopened.

When the normally closed stop button is pushed, input *B* of the AND gate changes from high to low. When input *B* changes to a low state, the output of the AND gate also changes to a low state. This causes a low to appear at input *B* of the OR gate. The OR gate now has both of its inputs low, so its output changes from a high state to a low state. Since input *A* of the AND gate is now low, the output is forced to remain low when the stop button returns to its closed position and input *B* becomes high. The circuit designed here can be turned on with the start button and turned off with the stop button.

The next design task is to connect the overload contact to the circuit. The overload contact must be connected in such a manner that it will cause the output of the circuit to turn off if it opens. One's first impulse might be to connect the overload contact to the circuit as shown in Figure 4–12B. In this circuit, the output of AND gate *#1* has been connected to input *A* of AND gate *#2*. Input *B* of AND gate *#2* has been connected to a high through the normally closed overload contact. If the overload contact remains closed, input *B* will remain high. The output of AND gate *#2* is, therefore, controlled by input *A*. If the output of AND gate *#1* changes to a high state, the output of AND gate *#2* will also change to a high state. If the output of AND gate *#1* becomes low, the output of AND gate *#2* will become low also.

If the output of AND gate *#2* is high and the overload contact opens, input *B* will become low and the output will change from a high to a low state. This circuit appears to operate with the same logic as the relay circuit until the logic is examined closely. Assume that the overload contacts are closed and the output of AND gate *#1* is high. Since both inputs of AND gate *#2* are high, the output is also high. Now assume that the overload contact opens and causes input *B* to change to a low condition. This forces the output of AND gate *#2* to change to a low state also. Input *A* of AND gate *#2* is still high, however. If the overload contact is reset, the output will immediately change back to a high state. If the overload contact opens and is then reset in the relay circuit, the relay will not restart itself. The start button must be pushed to restart the circuit. Although this is a small difference in circuit logic, it could become a safety hazard in some cases.

This fault can be corrected with a slight design change. Refer to Figure 4–12C. In this circuit, the normally closed stop button has been connected to input *A* of AND gate *#2*,

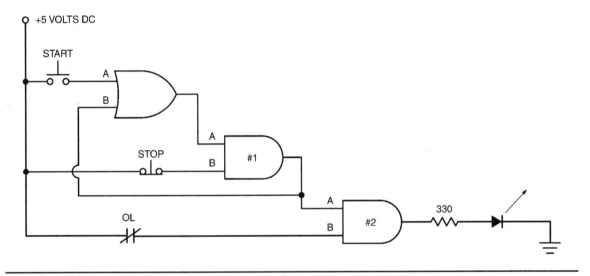

Figure 4–12B Digital-logic circuit.

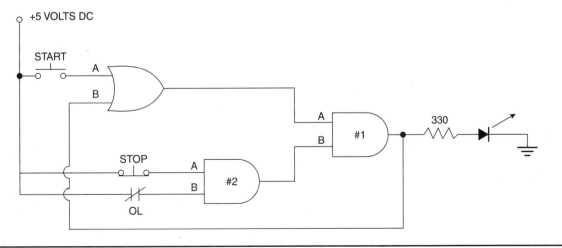

Figure 4–12C Logic schematic of start/stop circuit.

and the normally closed overload switch has been connected to input *B*. As long as both of these inputs are high, the output of AND gate #2 will provide a high to input *B* of AND gate #1. If either the stop button or the overload contact opens, the output of AND gate #2 will change to a low state. When input *B* of AND gate #2 changes to a low state, it will cause the output of AND gate #1 to change to a low state and unlock the circuit, just as pushing the stop button did in the circuit shown in Figure 4–12B. The logic of this digital circuit is now the same as the relay circuit.

Although the logic of this circuit is now correct, there are still some problems that must be corrected. When gates are used, their inputs must be connected to a definite high or low. When the start button is in its normal position, input *A* of the OR gate is not connected to anything. When an input is left in this condition, the gate may not be able to determine if the input should be high or low. The gate could, therefore, assume either condition. To prevent this, inputs must always be connected to a definite high or low.

When using TTL, inputs are always pulled high with a resistor as opposed to being pulled low. If a resistor is used to pull an input low as shown in Figure 4–13A, it will cause the gate to have a voltage drop at its output. This means that in the high state, the output of the gate may be only 3 or 4 volts instead of 5 volts. If this output is used as the input of another gate and the other gate has been pulled low with a resistor, the output of the second gate may be only 2 or 3 volts. Notice that each time a gate is pulled low through a resistor, its output voltage becomes low. If this were done through several steps, the output voltage would soon become so low that it could not be used to drive the input of another gate.

Figure 4–13B shows a resistor used to pull the input of a gate high. In this circuit, the push button is used to connect the input of the gate to ground, or low.

The push button can be adapted to produce a high at the input instead of a low by adding an INVERTER, as shown in Figure 4–13C. In this circuit, a pull-up resistor is connected to the input of an INVERTER. Since the input of the INVERTER is high, its output will produce a low at input *A* of the OR gate. When the normally open push button is pressed, a low will be produced at the input of the INVERTER. When the input of the INVERTER becomes low, its output becomes high. Notice that the push button will now produce a high at input *A* of the OR gate when it is pushed.

Since both of the push buttons and the normally closed overload contact are used to provide high inputs, the circuit is changed as shown in Figure 4–13D. Notice that the normally

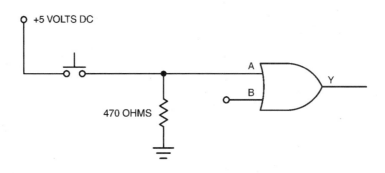

Figure 4–13A   Resistor used to lower the input of a gate.

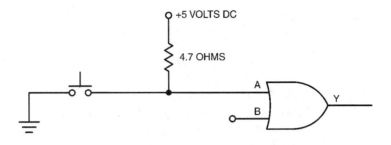

Figure 4–13B   Resistor used to raise the input of a gate.

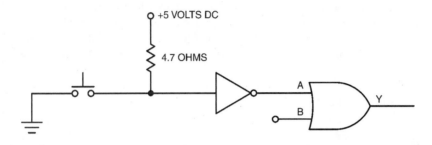

Figure 4–13C   Push button produces a high at the input.

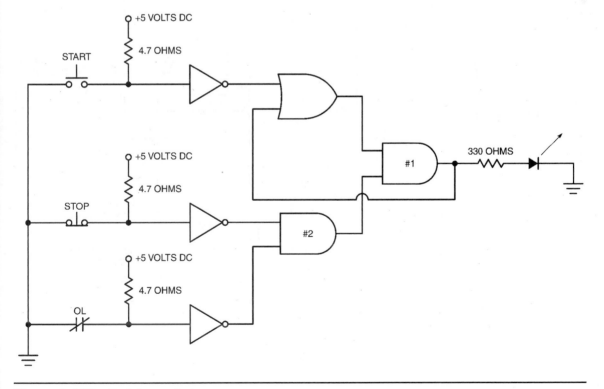

Figure 4–13D   Push-button logic.

closed push button and the normally closed overload switch connected to the inputs of AND gate *#2* are connected to ground instead of volts common collector (Vcc). When the switches are connected to ground, a low is provided to the input of the INVERTERS to which they are connected. The INVERTERS, therefore, produce a high at the input of the AND gate. If one of these normally closed switches opens, a high will be provided to the input of the INVERTER. This will cause the output of the INVERTER to become low. If the logic of the circuit shown in Figure 4–13A is checked, it can be seen that it is the same as the logic of the circuit in Figure 4–12C.

The final design problem for this circuit concerns the output. So far, a light-emitting diode has been used as the load. The LED is used to indicate when the output is high and when it is low. The original circuit, however, was used to control a 120-volt AC motor. This control can be accomplished by connecting a solid-state relay to the output in place of the LED (Figure 4–14). In this circuit, the output of AND gate *#1* is connected to the input of an opto-isolated, solid-state relay. When the output of the AND gate changes to a high condition, the solid-state relay turns on and connects the 120-volt AC load to the line.

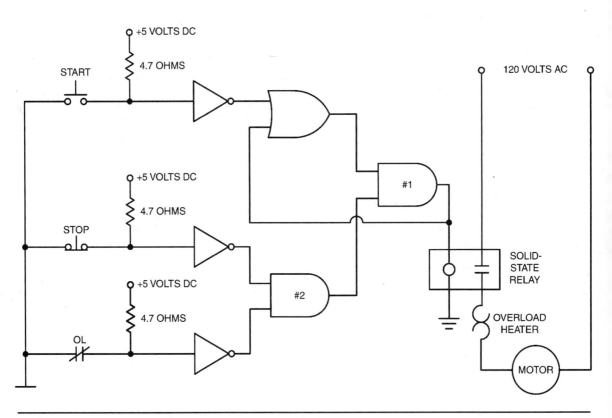

Figure 4–14 Solid-state, start/stop, push-button control.

CHAPTER 5

Starters

Manual Starters*

A *manual starter* is a motor controller whose contact mechanism is operated by a mechanical linkage from a hand-operated toggle or push button. A thermal unit and a direct acting overload mechanism provide motor overload protection. Basically, a manual starter is an on-off switch with overload relays. Manual starters are generally used on small machine tools, fans and blowers, pumps, compressors, and conveyors. They have the lowest cost of all motor starters, have a simple mechanism, and provide quiet operation with no AC hum. Moving a handle to ON or pushing the START button closes the power contacts, which remain closed until the handle is moved to OFF, or the STOP button is pushed, or the overload relay thermal units trip.

Manual starters are of the *fractional-horsepower type* or the *integral-horsepower type* and usually provide *across-the-line starting*. Standard manual starters cannot provide low-voltage protection or low-voltage release. If power fails, the contacts remain closed and the motor restarts when power returns. This is an advantage for devices such as pumps, fans, compressors, and oil burners, but for other applications it can be a disadvantage and can even be dangerous to personnel or equipment.

Manual starters or other maintained contact devices should not be used in applications where the machine or operator can be endangered if power fails and then returns without warning. For dangerous applications such as band saws, drill presses, or stamping presses, a magnetic starter and a momentary-contact pilot device giving three-wire control or a manual starter with low-voltage protection should be used for safety purposes.

Fractional-Horsepower Starters

Fractional-horsepower starters are designed to control and provide overload protection for motors of one horsepower or less on 115- or 230-volt single-phase supplies. They are available in single- and two-pole versions and are operated by a toggle handle on the front. When a serious overload occurs, the thermal unit trips to open the starter contacts, thus

*Adapted with permission from Square D Company.

disconnecting the motor from the line. The contacts cannot be reclosed until the overload relay has been reset by moving the handle to the OFF position.

Manual Motor-Starting Switches   *Manual-motor starting switches* provide on-off control of single-phase or three-phase AC motors where overload protection is not required or is separately provided. Two- or three-pole switches are available with ratings up to 10 horsepower, 600 volts, three-phase. The continuous current rating is 30 amperes at 250 volts maximum and 20 amperes at 600 volts maximum.

The toggle operation of the manual switch is similar to that of the fractional-horsepower starter. Typical applications of the switch include small machine tools, pumps, fans, conveyors, and other electric machinery that has separate motor overload protection. They are particularly suited to switch nonmotor loads, as in resistance heaters.

The two-pole starters have one overload relay, whereas three-pole starters usually have three overload relays. When an overload relay trips, the starter mechanism unlatches, thus opening the contacts to stop the motor. The contacts cannot be reclosed until the starter mechanism has been reset by pressing the STOP button or moving the handle to the RESET position after allowing time for the thermal unit to cool.

Integral-Horsepower Starters

Integral-horsepower starters with low-voltage protection prevent automatic restart of motors after a power loss. This is accomplished with a continuous-duty solenoid that is energized whenever the line-side voltage is present. If the line voltage is lost or disconnected, the solenoid deenergizes, thus opening the starter contacts. The contacts will not automatically close when the line voltage is restored. To close the contacts, the device must be manually reset and the START button must be pressed. This manual starter will not function unless the line terminals are energized. Typical applications include conveyors, grinders, mixers, metal and woodworking machinery, and applications where local or national electrical codes require low-voltage protection.

Magnetic Starters

A high percentage of applications require the controller to be capable of operation from remote locations or of providing automatic operation in response to signals from pilot devices such as thermostats, pressure or float switches, and limit switches. Low-voltage release or low-voltage protection might also be desired. Manual starters cannot provide this type of control; therefore, *magnetic starters* are used.

The operating principle that distinguishes a magnetic starter from a manual starter is the use of an *electromagnet*. The electromagnet consists of a *coil* of wire placed on an iron core. When current flows through the coil, the iron of the magnet becomes magnetized, attracting the iron bar called the *armature* (Figure 5–1). The electromagnet can be compared to a permanent magnet in that both will attract the iron bar.

The magnetic field of a *permanent* magnet will hold the armature against the pole faces of the magnet indefinitely. As a result, the armature cannot be dropped out except by physically pulling it away. In the electromagnet, however, interrupting the current flow through the coil of wire causes the magnet to become demagnetized and to drop out the armature.

A manual starter must be mounted so that it is easily accessible to the operator. However, a magnetic starter may be remotely mounted with the push-button stations or other pilot devices mounted anywhere on the machine and connected by control wiring into the coil circuit of the starter.

Magnetic Contactors

Most motor applications require the use of remote-control devices to start and stop the motor. *Magnetic contactors,* similar to the ones shown in Figure 5–1 and Figure 5–2, are commonly used to provide this function. Contactors are also used to control distribution of power in lighting and heating circuits.

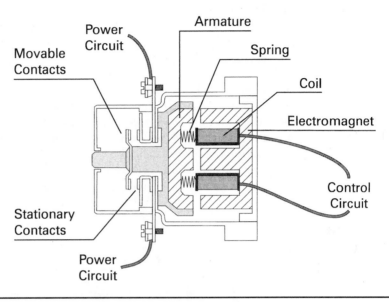

Figure 5–1   Internal parts of a contactor. Courtesy Siemens Energy and Automation Inc.

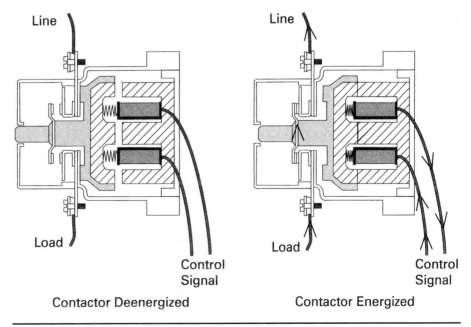

Figure 5–2   Contactor operation. Courtesy Siemens Energy and Automation Inc.

Both figures show the interiors of basic contactors. There are two circuits involved in the operation of a contact: the control circuit and the power circuit. The *control circuit* is connected to the coil of an electromagnet, and the *power circuit* is connected to the stationary contacts.

When power is supplied to the coil from the control circuit, a *magnetic field* is produced, magnetizing the electromagnet. The magnetic field attracts the armature, which has the movable contacts, to the stationary contacts. With the contacts closed, current flows through the power circuit from the line to the load. When the electromagnet's coil is deenergized, the magnetic field collapses and the movable contacts open under spring pressure. Current no longer flows through the power circuit.

The schematic in Figure 5–3 shows the electromagnetic coil of a contact connected to the control circuit through a switch (SW_1). The contacts are connected in the power circuit to the AC line and a three-phase motor. When SW_1 is closed, the electromagnetic coil is energized, closing the M contacts and applying power to the motor. Opening SW_1 deenergizes the coil, and the M contacts open, removing power from the motor.

Contactors are used to control power in a variety of applications. When applied in motor-control applications, contactors can only start and stop motors. They provide no overload protection. Most motor applications require overload protection. Overload relays similar to the one shown in Figure 5–4, provide this protection. The operating principle,

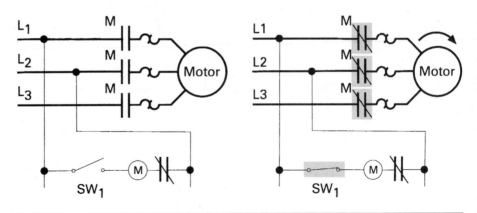

Figure 5–3 Wiring diagram and schematic of a two-wire manual start circuit. Courtesy Siemens Energy and Automation Inc.

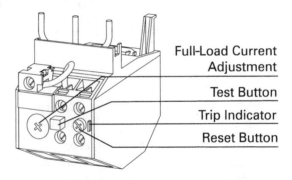

Figure 5–4 Overload relay. Courtesy Siemens Energy and Automation Inc.

using heaters and bimetal strips, is similar to that of the overloads in manual starters. Contactors and overload relays are separate control devices. When a contactor is combined with an overload relay, it is called a *motor starter.*

Anatomy of a Starter

One of the best ways to learn how to troubleshoot and design control systems is to understand the internal workings of the starter. Figure 5–5A shows the basic parts of the starter. Figure 5–5B gives a more detailed illustration of a starter, and the following text breaks down these parts in more detail.

Magnet and Armature Assemblies*

In the construction of a magnetic controller, the armature is mechanically connected to a set of contacts so that when the armature moves to its closed position, the contacts also close. Figure 5–6 shows several *magnet and armature assemblies* in elementary form. When the coil has been energized and the armature has moved to the closed position, the controller is said to be *picked up* and the armature *seated* or *sealed-in.*

Among the common types of magnet and armature assemblies are:

- *Clapper-type:* The armature is hinged. As it pivots to close (seal in), the movable contacts close against the stationary contacts.

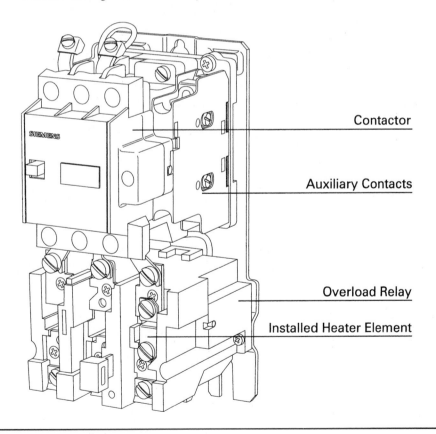

Figure 5–5A   A starter. Courtesy Siemens Energy and Automation Inc.

*Adapted with permission from Square D Company.

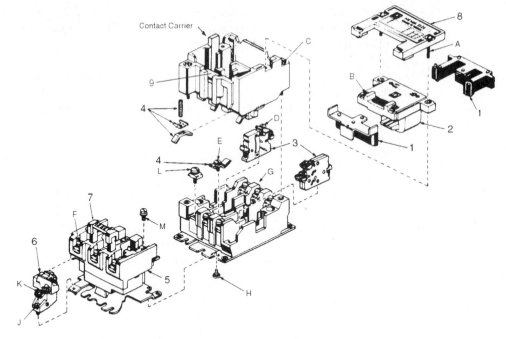

Key:

1. Armature and magnet kit
2. Coil
3. Internal auxiliary contact
   —Normally open
   —Normally closed
4. Contact kit
5. Melting-alloy-overload-relay assembly
   —One-element
   —Three-element
6. Melting-alloy-overload-contact unit
7. Reset bar
8. Cover
9. Lever bearing

A. Cover screw
B. Coil terminal-pressure wire connector
C. Power plant screw
D. Internal auxiliary contact
E. Contact kit
F. Overload-relay wire clamp and screw/Size 00 starter
G. Auxiliary wire binding screw
H. Screw
I.
J, K. Melting-alloy-overload-contact unit
L. Wire clamp and screw/Size 00 contact/Size 00 starter
M. Overload thermal unit fastening screw

Figure 5–5B Internal parts of a three-pole AC magnetic starter. Courtesy Square D Company.

- *Vertical-action:* The action is a straight-line motion with the armature and contacts being guided so that they move in a vertical plane.

- *Horizontal-action:* Both armature and contacts move in a straight line through a horizontal plane.

- *Bell-crank:* A bell-crank lever transforms the vertical action of the armature into a horizontal contact motion. The mechanical shock of the armature seating is not transmitted to the contacts, resulting in minimum contact bounce and longer contact life.

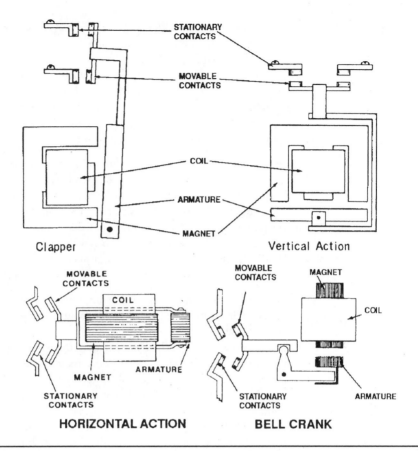

Figure 5–6   Magnet and armature assemblies. Courtesy Square D Company.

Magnetic Circuit

The *magnetic circuit* of a controller consists of the magnet assembly, the coil, and the armature. It is called a *circuit* from a comparison with an electric circuit. The coil and the current flowing in it cause magnetic flux to flow through the iron in a manner similar to a voltage causing current to flow through a system of conductors. The changing magnetic flux produced by AC results in a temperature rise in the iron parts. The heating effect is reduced by laminating the magnet assembly and armature.

Magnet Assembly   The *magnet assembly* is the *stationary* part of the magnetic circuit. The coil is supported by and surrounds part of the magnet assembly in order to induce magnetic flux into the magnetic circuit.

Armature   The *armature* is the *moving* part of the magnetic circuit. When it has been attracted into seated or sealed-in position, it completes the magnetic circuit. To provide maximum pull (to close the contacts) and to help ensure quietness, the faces of the armature and the magnet assembly are ground to a very close tolerance.

Air Gap   When a controller's armature has seated, it is held closely against the magnet assembly. However, a small gap is deliberately left in the iron circuit. When the coil is deenergized, some magnetic flux known as *residual magnetism* always remains, and, if it were not for the gap in the iron circuit, the residual magnetism might be sufficient to hold the armature in the sealed-in position.

An *air gap* can be created during the manufacturing process when the magnet faces are ground to close tolerances, and it is sometimes described as a *ground-air gap*. Some magnet assemblies contain a shim of precise thickness made of nonmagnetic metal placed in the magnetic flux path. This type of air gap is called a *permanent air gap*. By design, a magnet assembly could incorporate either or both types of air gaps.

Shading Coil   A *shading coil* is a single turn of conducting material (generally copper or aluminum) mounted in the face of the magnet assembly or armature (see Figure 5–7). The alternating main magnetic flux induces currents in the shading coil, and these currents set up auxiliary magnetic flux that is out of phase from the main flux. The auxiliary flux produces a magnetic pull that is out of phase from the pull due to the main flux, and this keeps the armature sealed in when the main flux falls to zero (which occurs 120 times per second with 60-cycle AC). Without the shading coil, the armature would tend to open each time the main flux goes through zero. Excessive noise, wear on the magnet faces, and heat would result.

Magnetic Coil   The *magnetic coil* has many turns of insulated copper wire wound on a spool. Most coils are protected by an epoxy molding, which makes them very resistant to mechanical damage.

When the controller is in the open position, there is a large air gap (not to be confused with the air gap in the magnetic circuit since the armature is at its farthest distance from the magnet. The *impedance* of the coil, which in AC magnetic circuits is the property to limit or resist current flow, is relatively low due to the air gap so that when the coil is energized it draws a fairly high current. As the armature moves closer to the magnet assembly, the air gap is progressively reduced and with it the coil current until the armature has sealed in. The final current is referred to as the *sealed current*. The *inrush current* is approximately six to ten times the sealed current; however, the ratio varies with individual designs. After the controller has been energized for some time, the coil becomes hot. This causes the coil current to fall to approximately 80 percent of its value when cold.

Alternating current magnetic coils should never be electrically connected in series. If one device seals in ahead of the other (which is quite likely if the devices are not identical and a possibility even if they are), the increased circuit impedance will reduce the coil cur-

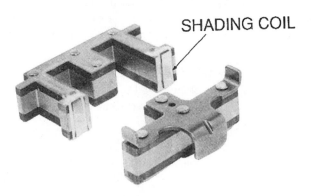

SHADING COIL

Figure 5–7   Magnet and armature assembly with shading coil shown. Courtesy Square D Company.

rent so that the slow device will not pick up or, if it does pick up, will not seal. For this reason, AC coils should always be connected in parallel.

Pick-Up Voltage/Seal-In Voltage   The *pick-up voltage* is the minimum voltage applied to the coil that causes the armature to move. The *seal-in voltage* is the minimum control voltage required to cause the armature to seat against the faces of the magnet. On devices using a vertical-action magnet and armature, the seal-in voltage is higher than the pick-up voltage to provide additional magnetic pull to ensure good contact pressure.

Control devices using the bell-crank armature and magnet arrangement are unique in that they have different force characteristics. Devices using this operating principle are designed to have a lower seal-in voltage than pick-up voltage. Contact life is extended, and contact damage under abnormal voltage conditions is reduced.

Drop-Out Voltage   If the control voltage is reduced sufficiently, the controller will open. The voltage at which this happens is called the *drop-out voltage* and is lower than the seal-in voltage.

Voltage Variation   NEMA standards require that the magnetic device operate properly at varying control voltages from a high of 110 percent to a low of 85 percent of rated coil voltage. This range of *voltage variation,* established by coil design, ensures that the coil will withstand given temperature rises at voltages up to 10 percent over rated voltage and that the armature will pick up and seal in even though the voltage may drop to 15 percent under the nominal rating.

If the voltage applied to the coil is *too high,* the coil will draw more than its designed current. As a result, excessive heat is produced and will cause early failure of the coil insulation. The magnetic pull will be too high and will cause the armature to seat with excessive force. The magnet faces will wear rapidly, leading to a shortened life for the controller. In addition, contact bounce may be excessive, resulting in reduced contact life.

Low control voltage produces low coil currents and reduced magnetic pull. On devices with vertical-action assemblies, if the voltage is greater than pick-up voltage but less than seal-in voltage, the controller may pick up but will not seal. With this condition, the coil current will not fall to the sealed value. As the coil is not designed to continuously carry a current greater than its sealed current, it will quickly get very hot and burn out. The armature will also *chatter.* In addition to the noise, wear on the magnet faces results.

In both vertical-action and bell-crank construction, if the armature does not seal, the contacts will not close with adequate pressure. Excessive heat, with arcing and possible welding of the contacts, will occur as the controller attempts to carry current with insufficient contact pressure.

Power Circuit

The *power circuit* of a starter includes the stationary and movable contacts and the thermal unit or heater portion of the overload relay assembly. The number of contacts (or power poles) is determined by the electric service. In a three-phase, three-wire system, for example, a three-pole starter is required (Figure 5–8).

NEMA Sizes and Ratings

Power-circuit contacts handle the motor load. The ability of the contacts to carry the full-load current without exceeding a rated temperature rise and their isolation from adjacent contacts correspond to NEMA standards established to categorize the NEMA size of the

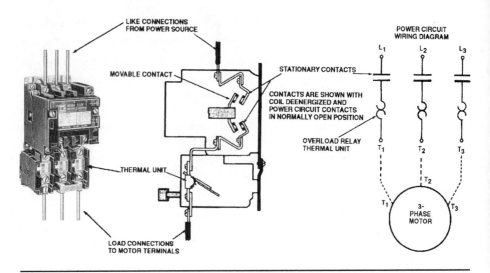

Figure 5–8 Magnetic starter power circuit. Courtesy Square D Company.

starter. The starter must also be capable of interrupting the motor circuit under locked-rotor-current conditions.

To be suitable for a given motor application, the magnetic starter selected should be equal to or exceed the motor horsepower and full-load current ratings. For example, a 50-horsepower motor with 230-volt service, polyphase, and a full-load current of 125 amperes will need a NEMA size 4 starter for normal motor duty. This same NEMA size 4 starter is derated to motors of 30-horsepower maximum if jogging or plugging duty is required. If the same 50-horsepower motor is used for jogging or plugging duty, the only controller rated for plugging or jogging duty for a 50-horsepower motor with a 230-volt load is a NEMA size 5 with a 75-horsepower rating.

Coil Circuit

The circuit to the magnetic coil, which causes a magnetic starter to pick up and drop out, is distinct from the power circuit. Although the power circuit can be single phase or polyphase, the *coil circuit* is always a single-phase circuit. Elements of a coil circuit (Figure 5–9) include the following:

- The magnet coil
- The control contact(s) of the overload relay assembly
- A momentary or maintained contact pilot device, such as a push-button station or a pressure, temperature, liquid level, or limit switch
- In lieu of a pilot device, the contact(s) of a relay or timer
- An auxiliary contact on the starter, designated as a holding-circuit interlock, which is required in certain control schemes

The coil circuit is generally identified as the *control circuit,* and contacts in the control circuit handle the coil load.

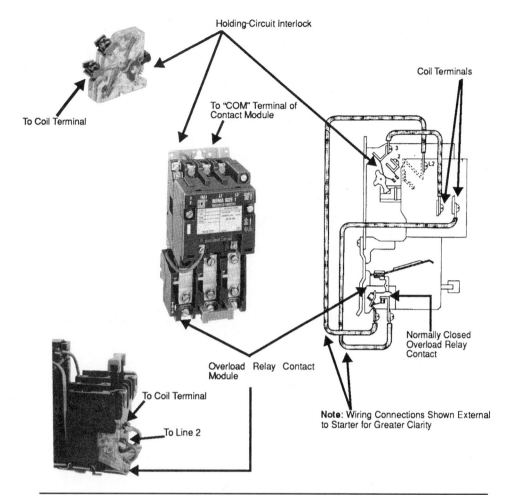

Holding-Circuit interlock

Coil Terminals

To Coil Terminal

To "COM" Terminal of
Contact Module

Normally Closed
Overload Relay
Contact

Overload Relay Contact
Module

To Coil Terminal

To Line 2

Note: Wiring Connections Shown External
to Starter for Greater Clarity

Figure 5–9   Coil circuit. Courtesy Square D Company.

Control Circuit   The interwiring shown in Figure 5–10 covers only the control-circuit wiring provided by the factory. Per NEMA standards, the single-phase control circuit is conventionally wired between line 1 (L_1) and line 2 (L_2). As review of the wiring diagram shows, the control circuit is connected to the single-phase circuit at L_2 and there is no direct control-circuit connection from line 1 (L_1) to line 3 at the holding-circuit interlock contact.

Control-Circuit Currents

Although the power-circuit and *control-circuit currents* may be the same, the voltage drawn by the motor in the power circuit is much higher than that drawn by the coil in the control circuit. Pilot devices and contacts of timers and relays used in control circuits are not generally horsepower-rated, and the current rating is low compared to the power circuits of starters or contactors. Inrush and sealed currents of a control circuit can be determined by reference to a magnetic coil table provided by the manufacturer.

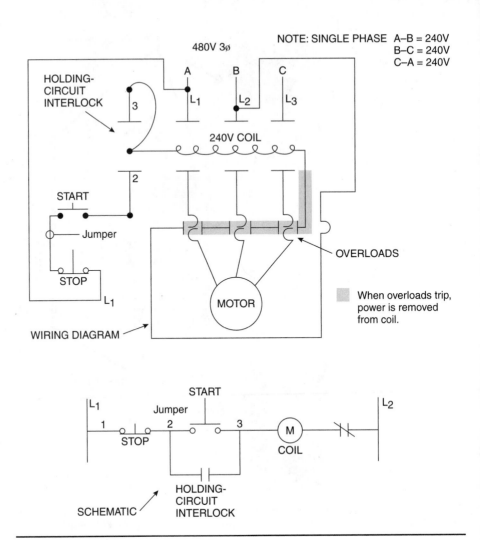

Figure 5-10 Magnetic starter control circuit.

A standard-duty push button with a rating of 15-ampere inrush, 1.5-ampere normal (sealed) with current at 240 volts 60 hertz can satisfactorily be used to control the coil circuit of a three-pole NEMA size 3 starter or contact, which has an inrush current of 2.9 amperes (700 volt-amperes/240 volts). As a comparison of the differences in current, the power-circuit contacts of the aforementioned starter may be controlling a 30-horsepower polyphase motor, drawing a full-load current of 78 amperes.

Alternating current magnetic coils are designed to operate on line voltages fluctuating as much as 15 percent below and 10 percent above nominal rating. Direct current coils have corresponding limits of 20 percent below and 10 percent above nominal rating.

Motor-Starting Current

When most motors start, they draw current in excess of the motor's full-load current rating. Motors are designed to tolerate this excess current for a short period of time. Many motors require six times (600 percent) the full-load current rating to start. Some newer, high-efficiency motors may require higher starting currents. As the motor accelerates to operating speed, the current drops off quickly. The time it takes for a motor to accelerate to operating speed depends on the operating characteristics of the motor and the driven load. A motor, for example, might require 600 percent of full-load current and take 8 seconds to reach operating speed. The overload device must be capable of allowing the motor to exceed its full-load rating for a short time. Otherwise, the motor will trip each time it is started.

Overload Relay Classes   Overload relays are divided into three *classes:* Class 10, Class 20, and Class 30. These classes define the length of time allowed for an overload before the overload relay trips. Class 10 provides the highest level of protection. This is usually sufficient time for the motor to reach full speed.

The *standard overload protection* is Class 20 with interchangeable heater elements. Heater elements must be sized for specific motor current.

- *Class 10:* 600 percent full-load amperes trip within 10 seconds.
- *Class 20:* 600 percent full-load amperes trip within 20 seconds.
- *Class 30:* 600 percent full load amps trip within 30 seconds.

Heater Element Selection   The heater application table (given by the manufacturer with the starter) indicates the range of full-load motor current to which a given heater element may be applied. Heater elements should be selected on the basis of the actual full-load current and service factor as shown on the motor nameplate. When the service factor of a motor is 1.15 or 1.25, the heater element is selected directly from the manufacturer's tables. If the same motor had a service factor of 1.0, or a maximum of 115 percent protection is desired, select one size smaller heater element.

Optional Overload Relay   Optionally, the blocks of starters can be interchangeable with other starters. See the manufacturer's specifications to be sure. Some also detect phase loss and unbalanced loads.

Auxiliary Contacts

Holding-Circuit Interlock*   The *holding-circuit interlock* (see Figure 5–9) is a normally open *auxiliary contact* provided on standard magnetic starters and contacts. It closes when the coil is energized to form a holding circuit for the starter after the start button has been released. For reasons of economy or design, an additional normally open power pole is sometimes used to perform the holding-circuit-interlock function.

Electric Interlocks   In addition to the main or power contacts that carry the motor current and the holding-circuit interlock, a starter can be provided with externally attached

*Adapted with permission from Square D Company.

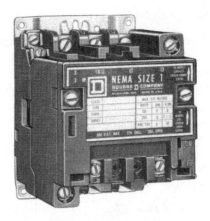

Figure 5–11 Magnetic contactor with detached external auxiliary contacts. Courtesy Square D Company.

auxiliary contacts, commonly called *electric interlocks* (Figure 5–11). Auxiliary contacts are rated to carry only control-circuit currents, not motor currents. Normally open and normally closed versions are available.

Among a wide variety of other applications, auxiliary contacts can be used to control magnetic devices where sequence operation is desired, to electrically prevent another controller from being energized at the same time and to make circuits indicate alarms, pilot lights, bells, or other signals. Auxiliary contacts are available in kit form and can easily be added in the field.

Control Devices

A device that is operated by some nonelectric means (such as the movement of a lever) and has contacts in the control circuit of a starter is called a *control device.* Operation of the control device controls the starter and hence the motor.

Typical control devices are control stations, limit switches, foot switches, pressure switches, and float switches. The control device may be of the maintained-contact or momentary-contact type. Some control devices have a horsepower rating and are used to directly control small motors through the operation of their contacts. When used in this way, separate overload protection such as a manual starter would be appropriate since the control device does not usually incorporate overload protection.

A *maintained-contact control device* is one which, when operated, will cause a set of contacts to open (or close) and stay open (or closed) until a deliberate reverse operation occurs. A conventional thermostat is a typical maintained-contact device. Maintained-contact control devices are used with two-wire control.

A standard push button is a typical *momentary-contact control device.* Pushing the button will cause normally open contacts to close and normally closed contacts to open. When the button is released, the contacts revert to their original states. Momentary-contact devices are used with three-wire control or jogging service.

Low-Voltage Release By the nature of its control-circuit connections, a two-wire control scheme provides *low-voltage release.* The term describes a condition in which a reduction or loss of voltage will stop the motor but in which motor operation will automatically resume as soon as power is restored.

If the two-wire control device is closed, a power failure or drop in voltage below the seal-in value will cause the starter to drop out. However, as soon as power is restored or the voltage returns to a level high enough to cause the armature to pick up and seal, the starter contacts will reclose and the motor will again run. This is an advantage in applications involving unattended pumps, refrigeration processes, and ventilating fans.

In many applications, however, the unexpected restarting of a motor after power failure is undesirable, as in a process where a number of motors must be restarted or operations performed in a prescribed sequence. In some applications, the automatic restart presents the possibility of danger to personnel or damage to machinery and work in process.

Low-Voltage Protection   If protection from the effects of a low-voltage condition is required, the two wire control scheme is not suitable. Three-wire control, which provides the desired protection, should be used. The three-wire control scheme provides *low-voltage protection.* In both two- and three-wire control, the starter drops out and the motor stops in response to a low-voltage condition or power failure.

When power is restored, however, the starter connected for three-wire control will not pick up because the reopened holding-circuit contact and the normally open start button contact prevent current flow to the coil. To restart the motor after a power failure, the low-voltage protection offered by three-wire control requires that the start button be depressed. A deliberate action must be performed, ensuring greater safety than that provided by two-wire control. Manual starters with low-voltage protection offer this same type of protection.

As the name implies, a full-voltage or across-the-line starter directly connects the motor to the power source. The starter can be either manual or magnetic. A motor connected in this fashion draws full inrush current and develops maximum starting torque so that it accelerates the load to full speed in the shortest possible time.

With some loads, the high starting torque will damage belts, gears and couplings, and material being processed. High inrush current can produce line voltage dips that cause lamp flicker and disturbances to other loads. Lower starting currents and torques are therefore often required and are achieved by reduced-voltage starting.

The coil circuit of a magnetic starter or contact is distinct from the power circuit. The coil circuit could be connected to any single-phase source of power, and the controller would be operable, provided the coil voltage and frequency match the service to which it is connected.

When the control circuit is connected to L_1 and L_2 of the starter, the voltage of the control circuit is always the same as the power-circuit voltage. The term *common control* is used to describe the relationship. Other variations include separate control and control through a control-circuit transformer.

It is sometimes desirable to operate push buttons or other control-circuit devices at a voltage lower than the motor voltage. In Figure 5–12, a single-phase control transformer (with dual voltage 240/480 volts primary, 120 volts secondary) has its 480-volt primary connected to the 480-volt three-phase, three-wire service brought into the starter.

Note, however, that the control circuit is now connected to the 120-volt secondary of the transformer, rather than being connected to L_1 and L_2, as in common control. The coil voltage is therefore 120 volts, and the push-button or other control devices operate at the same voltage level. *Fuses* are used to protect the control circuit, and it is common practice to ground one side of the transformer secondary.

Control of a power circuit by a lower control-circuit voltage can also be obtained by connecting the coil circuit to a separate control-voltage source, rather than to a transformer secondary. The term used to describe this wiring arrangement is *separate control.* As is evident

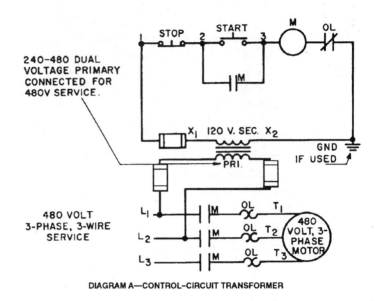

DIAGRAM A—CONTROL-CIRCUIT TRANSFORMER

Figure 5–12   Control-circuit transformer. Courtesy Square D Company.

from Figure 5–12, the coil rating must match the control-source voltage, but the power circuit can be any voltage (up to a 600-volt maximum).

Contactors

The general classification of *contactor* covers a type of electromagnetic device designed to handle relatively high currents. A special form of contactor exists for lighting load applications and is discussed separately.

The conventional *contactor* is identical in construction and current-carrying ability to the equivalent NEMA-size magnetic starter (see Table A–4 in Appendix A). The magnet assembly, coil, contacts, holding-circuit interlock, and other structural features are the same. The significant difference is that the contactor does not provide overload protection. Contactors, therefore, are used to switch high-current, nonmotor loads or are used in motor circuits if overload protection is separately provided. A typical application of the latter is in a reversing starter.

Standard Contactors   Filament-type lamps (tungsten, infrared, quartz) have inrush currents of approximately 15 to 17 times the normal operating currents. *Standard motor-control contactors* must be derated if used to control this type of load to prevent welding of the contacts on the high initial current.

A NEMA size 1 contactor has a continuous current rating of 27 amperes, but if used to switch certain lighting loads, it must be derated to 15 amperes. The standard contactor, however, need not be derated for resistance heating or fluorescent lamp loads since these do not impose as high an inrush current.

Lighting Contactors   *Lighting contactors* differ from standard contactors in that a holding-circuit interlock is not normally provided, since this type of contactor is frequently controlled by a two-wire pilot device such as a time clock or photoelectric relay. Unlike standard contactors, lighting contactors are not horsepower-rated or categorized by NEMA size but are designated by ampere ratings (20, 30, 60, 100, 200, and 300 amperes). It should be noted that lighting contactors are specialized in their application and should not be used on motor loads.

Electrically Held/Mechanically Held   In a conventional contactor, current flow through the coil creates a magnetic pull to seal in the armature and maintain the contacts in a switched position. Normally open contacts will be held closed; normally closed contacts will be held open. Because the contactor action is dependent on the current flow through the coil, the contactor is described as *electrically held*. As soon as the coil is deenergized, the contacts will revert to their initial position.

Mechanically held versions of contactors and relays are also available. The action is accomplished through the use of two coils and a latching mechanism. Energizing one coil, the *latch coil,* through a momentary signal causes the contacts to switch. A mechanical latch holds the contacts in this position even though the initiating signal is removed, and the coil is deenergized. To restore the contacts to their initial position, a second coil, the *unlatch coil,* is momentarily energized. Mechanically held contactors and relays are used where the slight hum of an electrically held device would be objectionable, as in auditoriums, hospitals, and churches.

Reversing Starter

Reversing the direction of motor shaft rotation is often required. Three-phase, squirrel-cage motors can be reversed by reconnecting any two of the three line connections to the motor. By interwiring the contactors, an electromagnetic method of making the reconnection can be obtained as seen in the power circuit in Figure 5–13. The contacts (*F*) of the forward contactor, when closed, connect L_1, L_2, and L_3 to motor terminals T_1, T_2, and T_3, respectively. As long as the forward contacts are closed, mechanical and electric interlock prevent the reverse contactor from being energized.

When the forward contactor is deenergized, the second contact can be picked up, closing its contacts that reconnect the lines to the motor. Note that by running through the reverse contacts, L_1 is connected to motor terminal T_3, and L_3 is connected to motor terminal T_1. The motor will now run in the opposite direction.

Whether operating through either the forward or reverse contactor, the power connections are run through an overload relay assembly, which provides motor overload protection. A *magnetic reversing starter,* therefore, consists of a starter and contactor that are suitably interwired with electric and mechanical interlocking to prevent the coil of both units from being energized at the same time.

Manual reversing starters employing two manual starters are also available. As in the magnetic version, the forward and reverse switching mechanisms are mechanically interlocked, but since coils are not used in the manually operated equipment, electric interlocks are not required.

Combination Starter and Disconnect

A *combination starter* is so named since it combines a magnetic starter and a *disconnect means,* which might incorporate a short-circuit protective device, in one enclosure. Compared with a separately mounted disconnect and starter, the combination starter takes up less space, requires less time to install and wire, and provides greater safety. Safety to personnel is assured because the door is mechanically interlocked so that it cannot be opened without first opening the disconnect. Combination starters can be furnished with circuit breakers or fuses to provide overcurrent protection and are available in nonreversing and reversing versions.

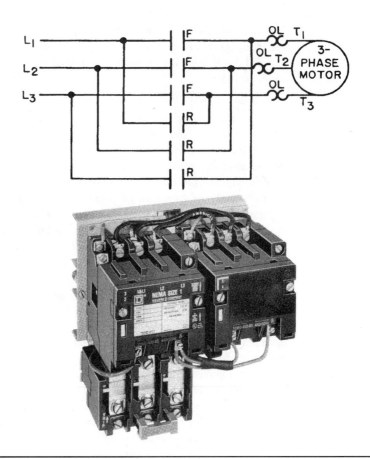

Figure 5–13   Three-pole reversing starter. Courtesy Square D Company.

Full-Voltage Starting

The most common type of motor starting is *full-voltage starting.* With this method, the motor is placed directly across the line. With this type of starter, the motor receives the full line voltage. When a motor is started with full voltage, starting current can be as high as 600 percent of full-load current on standard squirrel-cage motors. It can be even higher on high efficiency motors. Motor-starting torque is typically around 150 percent of full-load torque (see Figure 5–14).

Reduced-Voltage Starting

There are a number of reasons motors may need *reduced-voltage starting.* Many applications require the starting torque to be applied gradually. A conveyor belt, for example, requires the starting torque to be applied gradually to prevent belt slipping or bunching. Another reason reduced-voltage starting may be necessary is to reduce the effect a motor will have on the line when it starts. Transient overloading of the power supply caused by large motors starting can cause spikes or power surges that affect other connected equipment. Many power companies in the United States require reduced-voltage starting on large-horsepower motors. In general, starting methods that deviate from full-voltage starting by providing a lower starting voltage are referred to as reduced-voltage starting.

Reduced-voltage starting reduces the starting voltage of an induction motor with the purpose of confining the rate of change of the starting current to predetermined limits. It is important to remember that when the voltage is reduced to start a motor, current is also reduced, which also reduces the amount of starting torque a motor can deliver. Several methods are available for reduced-voltage starting. The application or the type of motor generally dictates the method to use.

Autotransformer Reduced-Voltage Starting

Autotransformer reduced-voltage starting provides the highest starting torque per ampere of line current and is one of the most effective means of starting a motor for an application in which starting current must be reduced with a minimum sacrifice of starting torque. Autotransformers have *adjustable taps* to reduce starting voltage to 50 percent, 65 percent, and 80 percent of full line voltage (see Figure 5–15).

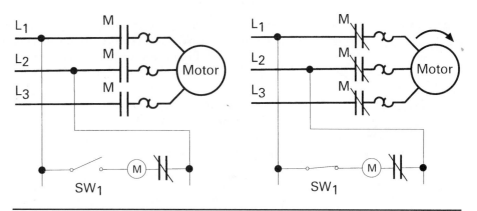

Figure 5–14 Full-voltage starting. Courtesy Siemens Energy and Automation Inc.

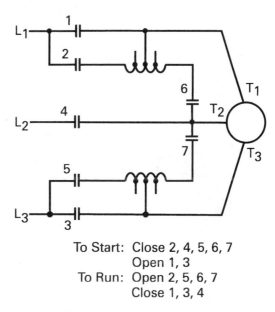

To Start: Close 2, 4, 5, 6, 7
Open 1, 3
To Run: Open 2, 5, 6, 7
Close 1, 3, 4

Figure 5–15 Autotransformer used for reduced-voltage starters. Courtesy Siemens Energy and Automation Inc.

Part-Winding Reduced-Voltage Starters

Part-winding reduced-voltage starters are used on motors with two separate parallel windings on the stator. The windings used during start draw about 65 to 80 percent of rated locked-rotor current. During run, each winding carries approximately 50 percent of the load current (see Figure 5–16).

Wye-Delta Reduced-Voltage Starters

Wye-delta reduced-voltage starters are applicable only with motors having stator windings not connected internally and all six motor leads available. Connected in a wye configuration, the motor starts with reduced starting line current. The motor is reconfigured to a delta connection for run (see Figure 5–17).

Solid-State Controllers

Solid-state controllers are available in 208 and 230 volts AC up to 150 horsepower and 460 volts AC up to 350 horsepower. The basic controller consists of electronic circuit boards and silicon-controlled rectifiers (SCR). This controller controls the firing sequence of the SCRs to provide smooth, soft starts and stops.

Multispeed Controllers

Full-voltage, AC, magnetic, *multispeed controllers* are designed to control squirrel-cage induction motors for operation at two, three, or four different constant speeds, depending on motor construction. The speed of a constant-speed motor is a function of the supply frequency and the number of poles and is given in the following formula:

Synchronous speed in rpm = 120 × frequency ÷ number of poles

The speed in rpm is the *synchronous speed* or the speed of the rotating magnetic field in the motor stator. Actual rotor speed is always less due to slip. The design of the motor and the amount of load applied determine the percentage of slip. This value is not the same for all motors. A motor with four poles on a 60-hertz AC line has a synchronous speed of

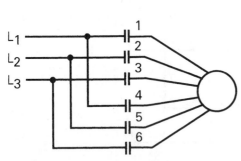

To Start: Close 1, 2, 3
          Open 4, 5, 6
To Run: Close 1, 2, 3, 4, 5, 6

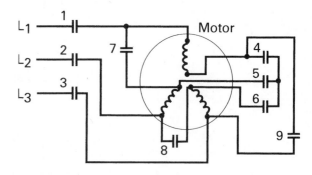

To Start: Close 1, 2, 3, 4, 5, 6
          Open 7, 8, 9
To Run: Open 4, 5, 6
          Close 7, 8, 9

Figure 5–16   Part-winding starter. Courtesy Siemens Energy and Automation Inc.

Figure 5–17   Wye-delta starter. Courtesy Siemens Energy and Automation Inc.

1800 rpm. This means that, after allowing for slip, the motor is likely to run at 1650 to 1750 rpm when loaded.

$$1800 = 120 \times 60 \div 4$$

An induction motor with two poles on a 60-hertz AC line, however, would run at twice that speed.

When a motor is required to run at different speeds, the motor's torque or horsepower characteristics will change with a change in speed. The motor must be properly selected and correctly connected for the application. In these applications, there are three categories:

1. Constant torque (CT)
2. Variable torque (VT)
3. Constant horsepower (CHP)

Separate-Winding Motors

There are two basic methods of providing multispeed control using magnetic starters: separate-winding motors and consequent-pole motors. *Separate-winding motors* have a separate winding for each speed. The motor cost is higher than for consequent-pole motors, but the control is simpler. There are many ways multispeed motors can be connected, depending on speed, torque, and horsepower requirements. The schematic in Figure 5–18 shows one possible connection of a two-speed, two-winding, wye-connected motor. The low-speed winding is wound for fewer poles than the high-speed winding.

Consequent-Pole Motors

Consequent-pole motors have a single winding for two speeds. Taps can be brought from the winding for reconnection for a different number of poles. Two-speed, consequent-pole motors have one reconnectable winding. Three-speed motors have one reconnectable winding and one fixed winding. Four-speed motors have two reconnectable windings.

Speed Selection

There are three control schemes of *speed selection* for multispeed motors: selective control, compelling control, and progressive control. *Selective control* permits motor starting at any

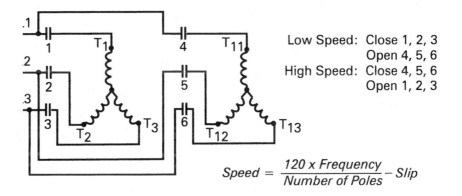

Low Speed: Close 1, 2, 3
Open 4, 5, 6
High Speed: Close 4, 5, 6
Open 1, 2, 3

$$Speed = \frac{120 \times Frequency}{Number\ of\ Poles} - Slip$$

Figure 5–18 Separate-winding starting. Courtesy Siemens Energy and Automation Inc.

speed. To move to a higher speed, the operator depresses the desired speed push button. *Compelling control* requires the motor to be started at the lowest speed; then the operator must manually progress in increments through each speed step to the desired speed. With *progressive control,* the motor is started at the lowest speed and automatically progresses in increments to the selected speed.

Reversing

Many applications require a motor to run in both directions. In order to change the direction of motor rotation, the direction of current flow through the windings must be changed. This is done on a three-phase motor by *reversing* any two of the three motor leads. Traditionally, T_1 and T_3 are reversed. Figure 5–19 shows a three-phase reversing motor circuit. It has one set of forward (*F*) contacts and one set of reverse (*R*) contacts. When the *F* contacts are closed, current flows through the motor causing it to turn in a clockwise direction.

When the *R* contacts are closed, current flows through the motor in the opposite direction causing it to rotate in a counterclockwise direction. Mechanical interlocks prevent both forward and reverse circuits from being energized at the same time.

Starter Ratings: NEMA and IEC

Starter contactors are rated according to the size and type of load they handle. Two organizations, the National Electrical Manufacturers Association (NEMA) and the International Electrotechnical Commission (IEC), rate contacts and motor starters. Though NEMA is primarily associated with equipment used in North America, IEC is associated with equipment sold in many countries, including the United States. International trade agreements, market globalization, and domestic and foreign competition have made it important for controls manufacturers to be increasingly aware of international standards.

There are differences between NEMA- and IEC-rated devices. Each organization has its own procedures and standards. The NEMA ratings are maximum horsepower ratings, according to the NEMA ICS2 standards. Sizes are specified by NEMA from size 00 to size 9; these sizes cover the horsepower range of 2 horsepower to 1600 horsepower at 460 volts AC. The IEC ratings are maximum operational current as specified by the IEC in publication IEC 158-1, Utilization Category AC-3. Because the IEC does not specify sizes, the buyer needs to make clear which standards he expects to be met.

There are several other organizations that have developed standards and tests for electric equipment. Underwriters Laboratory (UL), for example, specifies a maximum horsepower rating for which a contactor can be used. The contactor is tested by UL using test procedure UL 508.

Overload Relay and Full-Load Ampere Adjustment

The *overload relay* is a Class 10 relay that uses a bimetal strip unit and heater element to detect overloads. Each phase monitors current. The unit has a full-load ampere adjustment, test button, and reset button. The *full-load ampere adjustment* corresponds to the range of the motor full-load ampere rating. The test button is to ensure that the overload relay is functioning properly. The reset button is used to reset a trip. It can be either auto or manual. There is also a trip indicator (see Figure 5–20).

It is easy to adjust the overload relay for the proper trip point. For example, assume that the overload relay is being adjusted for a 10-horsepower, three-phase, 230-volt AC motor. From the nameplate of this example motor, it can be seen that the full-load current is 24 amperes, and the service factor is 1.15.

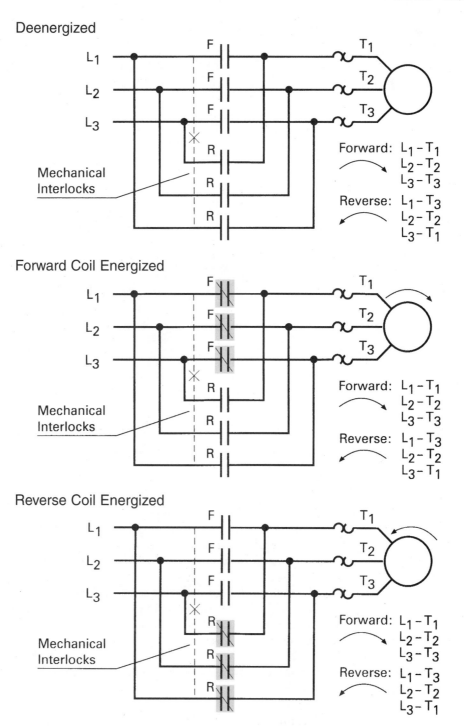

Figure 5–19 Reversing. Courtesy Siemens Energy and Automation Inc.

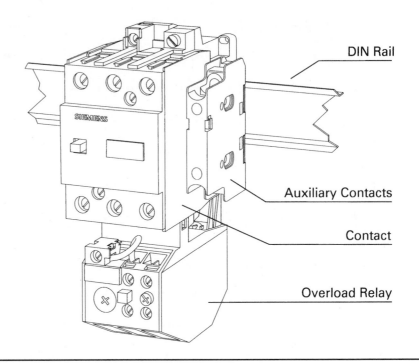

DIN Rail

Auxiliary Contacts

Contact

Overload Relay

Figure 5–20 Starter. Courtesy Siemens Energy and Automation Inc.

The trip point at which a motor overload relay must be set varies with motor design. The *NEC®* specifies that if a motor's service factor is 1.15 or more, or temperature rise is 40° C or less, the motor can be run on a current that is up to 125 percent of the motor's full-load amperes. The running current of all other motors cannot exceed 115 percent of the motor's full-load ampere rating. Some overload relays are factory-calibrated to trip at 125 percent of full-load ampere rating.

The overload will allow for motor inrush current (600 to 800 percent of full-load amperes). The full-load amperes trip adjustment is set to 24 amperes with this motor. The length of time an overload relay will allow a motor to remain energized varies with the amount of overload.

If the same motor had a nameplate service factor of 1.0, the *NEC®* would permit it to exceed its full-load ampere rating by 15 percent. To obtain the proper setting, multiply the full-load ampere rating (24) by .92 (Note: *1.25 × .92 = 1.15*). The correct setting would be approximately 22 amperes.

Connecting a Motor Starter

Figure 5–21 pictorially shows the connection of a contactor. Not all contactors are designed the same way as shown in the figure. This can be illustrated schematically with a line diagram. The control does not have to get its power from the same source as the motor. In this illustration, the control voltage is supplied from a separate 120-volt AC source. This is referred to as *separate control.*

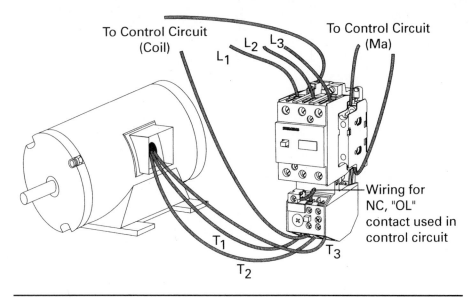

Figure 5–21   Starter connected to motor. Courtesy Siemens Energy and Automation Inc.

Trouble-shooting Starters

Alternating Current Hum

All AC devices that incorporate a magnetic effect produce a characteristic *hum*. This hum or noise is due mainly to the changing magnetic pull (as the flux changes) inducing mechanical vibrations. Excessive noise in the contactors, starters, and relays could result from the following operating conditions:

- Broken shading coil.
- Too-low operating voltage.
- Wrong coil.
- Misalignment between the armature and magnet assembly (the armature is then unable to seat properly).
- Dirt, rust, filings, etc., on the magnet faces (the armature is unable to seal-in completely).
- Jamming or binding of moving parts (contacts, springs, bearings) so that full travel of the armature is prevented.
- Incorrect mounting of the controller, for example, a thin piece of plywood fastened to a wall so that a sounding board effect is produced.

See Troubleshooting Chart, Appendix G.

Relays and Contactors

Relays and Their Uses

The *relay* is basically a communication carrier. It is used in the control of fluid power valves and in many machine sequence controls such as drilling, boring, milling, and grinding operations. Relays are also used as power amplifiers.

The relay is an electromechanical device. The major parts of the relay are shown in Figure 6–1A and Figure 6–1B. There are both stationary and moving contacts. The moving contacts are attached to the plunger. When the coil of the relay is energized, the plunger moves through the coil, closing the normally open contacts and opening the normally closed contacts. Figure 6–1A shows the contacts in the normal or deenergized condition of the coil.

When the coil of the relay is connected to a source of electric energy, the coil is energized. When the coil is energized, the plunger moves up as shown in Figure 6–1B, due to a force produced by a magnetic field. The distance that the plunger moves is generally short—about one-fourth inch or less.

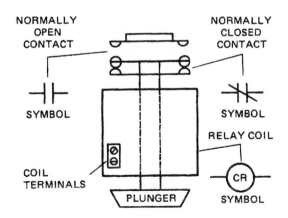

Figure 6–1A Deenergized condition of relay showing contact position.

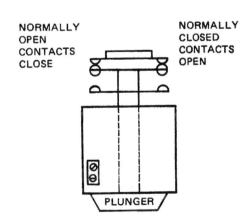

Figure 6–1B Energized condition of relay. With coil energized, magnetic force pulls the plunger up, operating the contacts.

There is a difference in the current in the relay coil from the time the coil is first energized and when the contacts are completely operated. When the coil is energized, the plunger is in an out position. Due to the open gap in the magnetic path (circuit), the initial current in the coil is high. The current level at this time is known as *inrush current*. As the plunger moves into the coil, closing the gap, the current level drops to a lower value. This lower value is called *sealed current*. The inrush current approximates six to eight times the sealed current.

Figure 6–1A also shows symbols for the *relay coil* and *contact*. One of the latest designs of control relays has electric clearances up to 600 volts. The contacts are convertible from open to closed, or from closed to open. Coils are available to cover most standard voltages from 24 volts to 600 volts. (The standard voltage used in machine control is 120 volts.) Relay coils are now being made of a molded construction. This aids in reducing moisture absorption and increases mechanical strength.

The level of voltage at which the relay coil is energized, resulting in the contacts moving from their normal, unoperated position to their operated position, is called *pick-up voltage*. After the relay is energized, the level of voltage on the relay coil at which the contacts return to their unoperated condition is called *drop-out voltage*.

Coils on electromechanical devices such as relays, contactors, and motor starters are designed so as not to drop out (deenergize) until the voltage drops to a minimum of 85 percent of the rated voltage. The relay coils also will not pick up (energize) until the voltage rises to 85 percent of the rated voltage. This voltage level is set by NEMA. Generally, 85 percent is found to be a conservative figure. Most electromechanical devices will not drop out until a lower voltage level is reached. Also, most electromechanical devices will pick up at a lower rising voltage level. Generally, coils on electromechanical devices will operate continuously at 110 percent of the rated voltage without damage to the coil.

The two important factors in a relay are the coil and the contacts. Of these, the contacts generally require greater consideration in practical circuit design. There are some single-break contacts used in industrial relays. However, most of the relays used in machine tool control have double-break contacts. The rating of contacts can be misleading. The three ratings generally published are:

1. *Inrush* or *make-contact* capacity
2. *Normal* or *continuous-carrying* capacity
3. The *opening* or *break* capacity

For example, a typical industrial relay has the following contact ratings:

- 10-ampere noninductive continuous load (AC)
- 6-ampere inductive load at 120 volts (AC)
- 60-ampere make and 60-ampere break, inductive load at 120 volts (AC)

The point to remember is that in determining the contact rating, it must be clear what rating is given.

A resistance is an example of a noninductive load. This may be a resistance unit used as a heating element. An inductive load is basically a coil. This could be a solenoid, contact coil, or motor-starter coil.

Relay contacts are usually silver or a silver-cadmium alloy. This material is used because of the excellent conductivity of silver. Silver oxide, which forms on the contacts, is also a good conductor. Adding a small amount of cadmium to the silver increases the interrupting capacity of the contacts. Some manufacturers now offer gold-plated contacts. This improves shelf life and provides improved contact reliability in low-energy, low-duty-cycle applications. There are several points that the circuit designer should consider in the use of control relays:

1. *Changing contacts from normally open to normally closed, or vice versa.* Most machine tool relays have some means of making this change, ranging from a simple flip-over contact to removing the contacts and relocating with spring change.

2. *Universal contacts.* This refers to relays where the contacts may be normally open or normally closed. Only one type may be selected in the use of each contact, but not both types.

3. *Split, or bifurcated, contacts.* As the name implies, the contact is divided into two parts. This provides for double the contact make points, improving the reliability of the contact and reducing contact bounce.

4. *Contact bounce.* All contacts bounce on closing. The problem is to reduce the bounce to a minimum. In rapid-operating relays, contact bounce can be a source of trouble. For example, an interlock in a transfer circuit can be lost by a contact opening momentarily. (The use of split contacts, overlapping contacts, other types of antibounce contacts, or a circuit change may help.)

5. *Overlap contacts.* In this case, one contact can be arranged to operate at a different time relative to another contact on the same relay. For example, the normally open contact can be arranged to close before the normally closed contact opens. This is called *make before break.*

6. *Contact wipe.* This results from the relative motion of the two contact surfaces after they make contact. That is, one contact wipes against the other. One advantage to be gained from this action is that it tends to provide a self-cleaning action for the contacts.

Panel mounting space has been a problem for the user of relays. With increasing complexity of control, the number of relays used has grown. This requires more panel space. Several years ago, a line of 300-volt relays was developed. The size was reduced from the old 600-volt relay. This change, along with a modular form of construction, was very helpful. More recently, a new line of 600-volt relays was introduced. These relays are comparable to the 300-volt line. The reference to 300-volt and 600-volt relays applies to the voltage being carried by the contacts. The standard relay coil voltage in the machine control field is 120 volts.

These relays are generally available with up to twelve contacts in combinations of normally open and normally closed contacts. They may be either convertible or fixed. In most cases, the relays can be combined with a pneumatic timing head or mechanical latch. However, when adding the pneumatic timing head, the relay is restricted to four instantaneous contacts.

Figure 6–2 shows a circuit diagram using a relay with two normally open contacts. One contact, *1CR$_1$*, is used as an interlock around the *START* push button. Thus, an *interlock circuit* is a path provided by electric energy to the load after the initial path has been opened. The second relay contact, *1CR$_2$*, is used to energize a light. Remember that when a relay coil is energized, the normally open contacts close. The circuit can be deenergized by operating the *STOP* push-button switch.

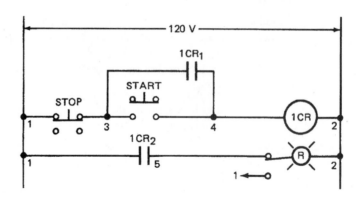

Figure 6–2 Control circuit.

Latching Relays

A mechanical latching attachment can be installed on the control relay (see Figure 6–3A). This obtains an interesting variation of the control capabilities of the relay. The *latching relay* is electromagnetically operated. It is held by means of a mechanical latch. By energizing a coil of the relay, called the *latch coil,* the relay operates. This results in the relay normally open contacts closing and the normally closed contacts opening. The electric energy can now be removed from the coil of the relay (deenergized), and the contacts remain in their operated condition. Now a second coil on the relay, the unlatch coil, must be energized in order to return the contacts to their unoperated condition. This arrangement is often referred to as a *memory relay.*

The latching relay has several advantages in electric circuit design. For example, it may be necessary to open or close contacts early in a cycle. At the same time, it may also be desirable to deenergize the section of the circuit responsible for the initial energizing of the relay latch coil. Later in the cycle, the unlatch coil can be energized to return the contacts to their original or unoperated condition. The circuit is then set up for the next cycle.

To further explain the relay:

1. Contacts should be shown in the unlatched condition, that is, as if the unlatch coil were the last one energized.

2. The latch and unlatch coils are on the same relay and always have the same reference number.

3. The contacts are always associated with the latch coil as far as reference designations are concerned.

Another use for the latching-type relay involves power failure. Here it may be necessary that the contacts remain in their operated condition during the power-off period. In this case, conditions are the same after the power failure as they were before.

If quietness of operation is desired in a long cycle, this can be provided. The coil can be deenergized, thus eliminating the usual hum. The conventional two-coil circuit for the latching relay is shown in Figure 6–3B. The latch coil of a latching relay replaces the coil of control relay *2CR.*

In Figure 6–3C, the closing of timing contact *2TR* energizes the latch coil *LCR.* Normally open contact *LCR,* closes, energizing a red pilot light. When the timing relay *TR*

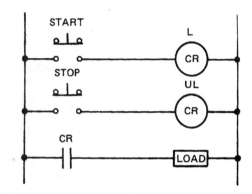

Figure 6–3A   Alternating current latching relay. Courtesy Square D Company.

Figure 6–3B   Control circuit.

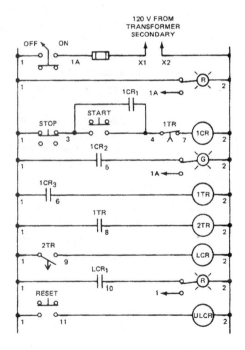

Figure 6–3C   Control circuit.

times out and the timing contact opens, the latch coil is deenergized. However, the light remains energized until the *RESET* push-button switch is operated, energizing the unlatch coil *ULCR*.

Plug-In Relays

Some industrial machine relays are available in plug-in types. *Plug-in relays* are designed for multiple switching applications at or below 240 volts. Coil voltages span standard levels from 6 volts to 120 volts. They are available for AC or DC. Mounting can be obtained with a tube-type socket, square-base socket mounting, or flange mounting using slip-on connectors.

The plug-in relay has a distinct advantage for changing relays without disturbing the circuit wiring. In critical operations in which the relay service is very hard and downtime is a premium, the plug-in relay may have some advantages.

Control Relays

A *control relay* is an electromagnetic device similar in operating characteristics to a contact. The contact, however, is generally employed to switch power circuits or relatively high current loads. Relays, with few exceptions, are used in control circuits; consequently, their lower ratings (6 amperes maximum at 600 volts) reflect the reduced current levels at which they operate.

Though contactors generally have from one to five poles, it is not uncommon to find relays used in applications requiring ten or twelve poles per device. Although normally open and normally closed contacts can be provided, the great majority of contact applications use the normally open contact configuration, and there is no conversion of contact operation in the field. Relays can use various combinations of normally open and normally closed contacts. In addition, some relays have convertible contacts, permitting changes to be made in the field from normally open to normally closed operation, or vice versa, without requiring kits or additional components.

Relays are used in control circuits of magnetic starters, contacts, solenoids, timers, and other relays. Relays are generally used to amplify the contact capability or multiply the switching functions of a pilot device.

Control-Relay Operation

When power is applied from the control circuit, an electromagnetic coil is energized. The resultant electromagnetic field pulls the armature and movable contacts toward the electromagnet, closing the contacts. When power is removed, spring tension pushes the armature and movable contacts away from the electromagnet, opening the contacts.

Contact Arrangement in Relays

A relay can contain normally open, normally closed, or both types of contacts. The main difference between a control relay and a contact is the size and number of contacts. The contacts in a control relay are relatively small because they need to handle only the small currents used in control circuits. There are no power contacts. Some relays have a greater number of contacts than are found in the typical contact. The use of contacts in relays can be complex. There are three words that must be understood when dealing with relays.

Pole   *Pole* refers to the number of isolated circuits that can pass through the relay at one time. A single-pole relay can carry current through one circuit. A double-pole or two-pole relay can carry current through two circuits simultaneously. The two circuits are mechanically connected so that they open or close at the same time.

Throw   *Throw* refers to the number of different closed contact positions per pole. This is the total number of different circuits each pole controls.

Break   *Break* refers to the number of separate contacts the switch contacts use to open or close individual circuits. If the switch breaks the circuit in one place, it is a single break. If the relay breaks the circuit in two places, it is a double break.

Interposing Relay

Figure 6–4 illustrates one way a control relay might be used in a circuit. A 24-volt AC coil may not be strong enough to operate a large starter. In this example the electromagnetic coil of the *M* contact selected is rated for 460 volts AC. The electromagnetic coil of the control relay *CR* selected is 24 volts AC. This is known as *interposing* a relay.

When the start button is momentarily depressed, power is supplied to the control relay *CR*. This energizes the control relay *CR;* closing the *CR* contacts in L_1 and L_2. The *CR* contacts in L_2 maintain the start circuit. The *CR* contacts in L_1 complete the path of current to the motor starter *M*. The motor starter *M* energizes and closes the *M* contacts in

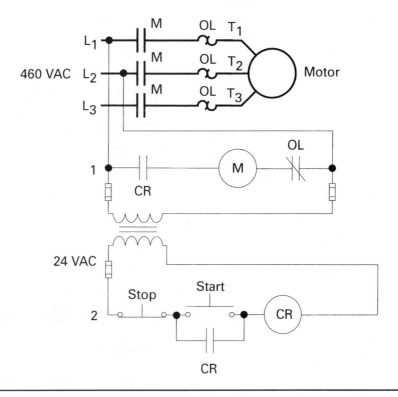

Figure 6–4   Interposing relay. Courtesy Siemens Energy and Automation, Inc.

the power circuit, starting the motor. Depressing the *Stop* push button deenergizes the control relay *CR* and motor starter *M*.

Voltage Amplification

Figure 6–5A represents a *voltage amplification.* A condition may exist in which the voltage rating of the temperature switch is too low to permit its direct use in a starter control circuit operating at some higher voltage. In this application, the coil of the interposing relay (*CR*) and the pilot device are wired to a low-voltage source of power compatible with the rating of the pilot device. The relay contact (*CR*), with its higher voltage rating, is then used to control the operation of the starter. When the temperature switch closes, the control relay is energized and normally open contacts *CR* are closed and energize coil *M*, which closes normally open contacts that feed the motor. When the temperature switch closes, the control relay is energized and normally open contacts *CR* are closed and energize coil *M*, which closes normally open contacts that feed the motor.

Figure 6–5B represents another use of relays, which is to multiply the switching functions of a pilot device with a single or limited number of contacts. Depressing the *ON* button in the control circuit energizes the relay coil (*CR*). Its normally open contacts close to complete the control circuits to the starter, solenoid, and timing relay, and one contact forms a holding circuit around the *ON* button. The normally closed contacts open to deenergize the contact and turn off the pilot light.

In the circuit shown in Figure 6–5B, a single-pole push-button contact can, through the use of an interposing six-pole relay, control the operation of a number of different loads such as a pilot light, starter, contact, solenoid, and timing relay.

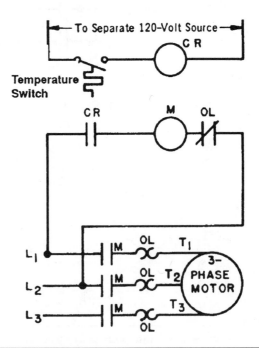

Figure 6–5A   Voltage amplification. Courtesy Square D Company.

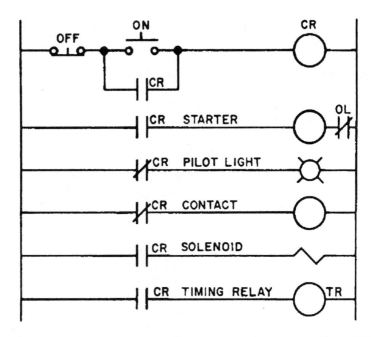

Figure 6–5B   Control circuit.

Relays are commonly used in complex controllers to provide the logic to set up and initiate the proper sequencing and control of a number of interrelated operations. Relays differ in voltage ratings (120, 300, 600 volts), number of contacts, contact convertibility, physical size, and in attachments to provide accessory functions such as mechanical latching and timing.

In selecting a relay for a particular application, one of the first steps should be to determine the control voltage at which the relay will operate. Once the voltage is known, the relays that have the necessary contact rating can be further reviewed and a selection made on the basis of the number of contacts and other characteristics needed.

Sequence of Operation

Referring to Figure 6–5B, when the *ON* switch is closed, coil *CR* is energized and the interlocking contact paralleled around the *ON* switch keeps the coil *CR* energized until the *OFF* switch is opened. When coil *CR* is energized, the following occurs:

1. The normally open *CR* contact for the starter is closed, and the starter coil is energized.

2. The normally closed *CR* contact for the pilot light is open and removes power to the light.

3. The normally closed *CR* contact for the contact is opened and the contact is deenergized.

4. The normally opened contact *CR* for the solenoid is closed and energizes the solenoid.

5. The normally open contact *CR* for the timing relay is closed and energizes the timing-relay coil.

Timer Relays

Timer relays are similar to other control relays in that they use a coil to control the operation of some number of contacts. The difference between a control relay and a timer relay is that the contacts of the timer relay delay changing their position when the coil is energized or deenergized.

Time-Delay Relays

Time-delay relays can be divided into two general classifications: the on-delay relay, and the off-delay relay. The *on-delay relay* is often referred to as *DOE*, which stands for *Delay On Energize*. The *off-delay relay* is often referred to as *DODE*, which stands for *Delay On DeEnergize*.

On-Delay Relay   When power is connected to the coil of an *on-delay relay,* the contacts delay changing position for some period of time. For this example, assume that the timer has been set for a delay of 10 seconds. Also assume that the contact is normally open. When voltage is connected to the coil of the on-delay relay, the contacts will remain in the open position for 10 seconds and then close. When voltage is removed and the coil is deenergized, contact will immediately change back to its normally open position. For contact symbols for the on-delay relay (see Figure 6–6).

Off-Delay Relay   The operation of the *off-delay relay* is the opposite of the operation of the on-delay relay. For this example, again assume that the timer has been set for a delay of 10 seconds, and also assume that the contact is normally open. When voltage is applied to the coil of the off-delay relay, the contact will change immediately from open to closed. When the coil is deenergized, however, the contact will remain in the closed position for 10 seconds before it reopens. The contact symbols for an off-delay relay are shown in Figure 6–6. Time-delay relays can have normally open, normally closed, or a combination of normally open and normally closed contacts.

Although the contact symbols shown in Figure 6–6 are standard NEMA symbols for on-delay and off-delay contacts, some control schematics may use a different method of indicating *timed contacts*. The abbreviations *TO* and *TC* are used with some control schematics to indicate a time-operated contact. *TO* stands for *time opening*, and *TC* stands

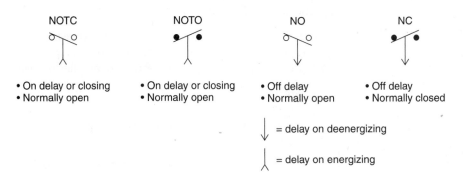

Figure 6–6   Time contact symbols.

for *time closing* (Figure 6–6 and Appendix E). If these abbreviations are used with standard contact symbols, their meaning can be confusing.

Dashpot Timers

Although timers are divided into two basic classifications, the on-delay and the off-delay, several methods are used to obtain these time delays. The *dashpot timer* operates by forcing a fluid to flow through orifices in a piston. The operation of a dashpot timer is the same as the operation of the dashpot overload. The only real difference between the dashpot timer and the dashpot overload is the type of coil used. The dashpot timer uses a voltage-operated coil, and the dashpot overload uses a current-operated coil.

Pneumatic Timers

Pneumatic or *air timers* operate by restricting the flow of air through an orifice to a rubber bellows or diaphragm. Figure 6–7 illustrates the principle of operation of a simple bellows timer. If rod *A* pushes against the end of the bellows, air is forced out of the bellows through the check valve as the bellows contracts. When the bellows is moved back, contact *TR* changes from an open to a closed contact. When rod *A* is pulled away from the bellows, the spring tries to return the bellows to its original position. Before the bellows can be returned to its original position, however, air must enter the bellows through the air inlet. The rate at which the air is permitted to enter the bellows is controlled by the needle valve. When the bellows returns to its original position, contact *TR* returns to its normally open position.

Pneumatic timers are popular throughout industry because they have the following characteristics:

1. They are unaffected by variations in ambient temperature or atmospheric pressure.

2. They are adjustable over a wide range of time periods.

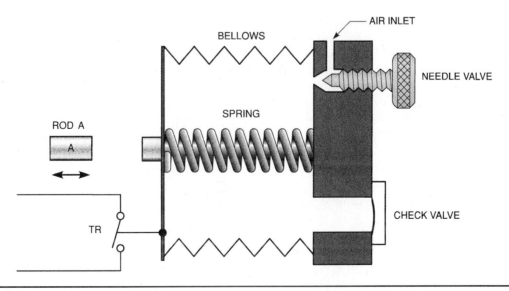

Figure 6–7   Bellows-operated pneumatic timer.

3. They have good repeat accuracy.

4. They are available with a variety of contact and timing arrangements.

Some pneumatic timers are designed to permit the timer to be changed from on delay to off delay and the contact arrangement to be changed to normally open or normally closed. This type of flexibility is another reason for the popularity of pneumatic timers.

Many timers are made with contacts that operate with the coil as well as time-delayed contacts. When these contacts are used, they are generally referred to as *instantaneous contacts* and are indicated on a schematic diagram by the abbreviation *inst.* printed below the contact. These instantaneous contacts change their positions immediately when the coil is energized and change back to their normal positions immediately when the coil is deenergized.

Clock Timers

Another timer frequently used is the *clock timer*. Clock timers use a small, AC, synchronous motor similar to the motor found in a wall clock to provide the time measurement for the timer. The length of time one clock timer measures may vary greatly from the length of time another measures. For example, one timer may have a full range of 0 to 5 seconds and another timer may have a full range of 0 to 5 hours. The same type of timer motor could be used with both timers. The gear ratio connected to the motor would determine the full range of time for the timer. Some advantages of clock timers are:

1. They have extremely high repeat accuracy.

2. Readjustment of the time setting is simple and can be done quickly. Clock timers are generally used when the machine operator must make adjustments of the time length.

Motor-Driven Timers

When a process has a definite on-and-off operation or a sequence of successive operations, a *motor-driven timer* is generally used. A typical application of a motor-driven timer is to control laundry washers where the loaded motor is run for a given period in one direction, reversed, and then run in the opposite direction.

Generally, this type of timer consists of a small, synchronous motor driving a cam-dial assembly on a common shaft. A motor-driven timer successively closes and opens switch contacts that are wired in circuits to energize control relays or contactors to achieve desired operations.

Capacitor Time-Limit Relay

Assume that a capacitor is charged by connecting it momentarily across a DC line and that the capacitor DC is then discharged through a relay coil. The current induced in the coil will decay slowly, depending on the relative values of capacitance, inductance, and resistance in the discharge circuit.

If a relay coil and a capacitor are connected parallel to a DC line (Figure 6–8), the capacitor is charged to the value of the line voltage and a current appears in the coil. If the coil-and-capacitor combination is now removed from the line, the current in the coil will start to decrease along the curve shown in Figure 6–8.

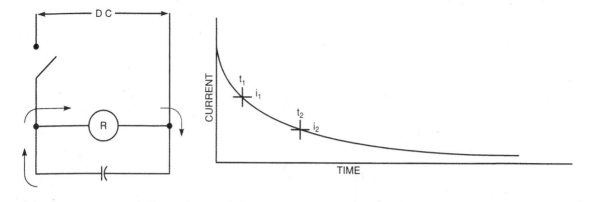

Figure 6–8 Charged capacitor discharging through a relay coil. The graph at the right illustrates the current decrease in the coil.

If the relay is adjusted so that the armature is released at current i_1, a time delay of t_1 is obtained. The time delay can be increased to a value of t_2 by adjusting the relay so that the armature will not be released until the current is reduced to a value of i_2.

A *potentiometer* is used as an adjustable resistor to vary the time. This resistance-capacitance (RC) theory is also used in industrial electronic and solid-state controls. This timer is highly accurate and is used in motor acceleration control and in many industrial processes.

Electronic Timers

Electronic timers use solid-state components to provide the time delay desired. Some of these timers use a resistance-capacitance time constant to obtain the time base, and others use quartz clocks as the time base. Resistance-capacitance time constants are inexpensive and have good repeat times. The quartz timers, however, are extremely accurate and can often be set for .1-second times. These timers are generally housed in a plastic case and are designed to be plugged into some type of socket. An electronic timer that is designed to be plugged into a standard eight-pin tube socket is shown in. The length of the time delay can be set by adjusting the control knob on top of the timer.

The schematic for a simple on-delay timer is shown in Figure 6–9. The timer operates as follows: When switch S_1 is closed, current flows through resistor RT and begins charging capacitor C_1. When capacitor C_1 has been charged to the trigger value of the injunction transistor, the uni-junction transistor (UJT) turns on and discharges capacitor C_1 through resistor R_2 to ground. The sudden discharge of capacitor C_1 causes a voltage spike to appear across resistor R_2. This voltage spike travels through capacitor C_2 and fires the gate of the SCR. When the SCR turns on, current is provided to the coil of relay K_1.

Resistor R_1 limits the current flow through the UJT. Resistor R_3 is used to keep the SCR turned off until the UJT provides the pulse to fire the gate. Diode D_1 is used to protect the circuit from the voltage spike produced by the collapsing magnetic field around coil K_1 when the current is turned off.

By adjusting resistor RT, capacitor C_1 can be charged at different rates. In this manner, the relay can be adjusted for time. Once the SCR has turned on, it will remain on until switch S_1 is opened.

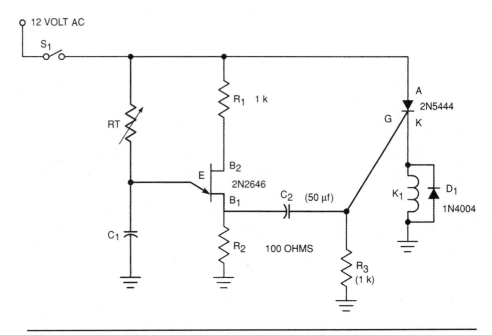

Figure 6-9 Schematic of electronic on-delay timer.

Programmable controllers (PCs) contain *internal* electronic timers. Most programmable controllers (PCs) use a quartz-operated clock as the time base. When the controller is programmed, the timers can be set in time increments of .1 second. This, of course, provides very accurate time delays for the controller.

Time-Controlled Circuit Application

Timed Starting for Three Motors

A machine contains three large motors. The current surge to start all three motors at the same time is too great for the system. Therefore, when the machine is to be started, there must be a delay of 10 seconds between the starting of each motor. The circuit shown in Figure 6–10A is a start-stop, push-button control that controls three motor starters and two time-delay relays. The circuit is designed so that an overload on any motor will stop all motors.

When the start button is pressed, a circuit is completed through the start button, motor starter coil M_1, and relay coil TR_1. When coil M_1 energizes, motor #1 starts and auxiliary contact M_1, which is parallel to the start button, closes. This contact maintains the current flow through the circuit when the start button is released (Figure 6–10B).

After a 10-second interval, contact TR_1 closes. When this contact closes, a circuit is completed through motor starter coil M_2 and timer relay coil TR_2. When coil M_2 energizes, motor #2 starts (Figure 6–10C).

Ten seconds after coil TR_2 energizes, contact TR_2 closes. When this contact closes, a circuit is completed to motor starter coil M_3, which causes motor #3 to start (Figure 6–10D).

If the stop button is pressed, the circuit to coils M_1 and TR_1 is broken. When motor starter M_1 deenergizes, motor #1 stops and auxiliary contact M_1 opens. TR_1 is an on-delay relay; therefore, when coil TR_1 is deenergized, contact TR_1 opens immediately.

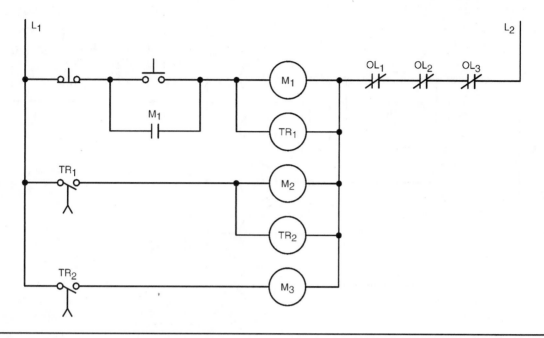

Figure 6–10A   Time-delay starting for three motors.

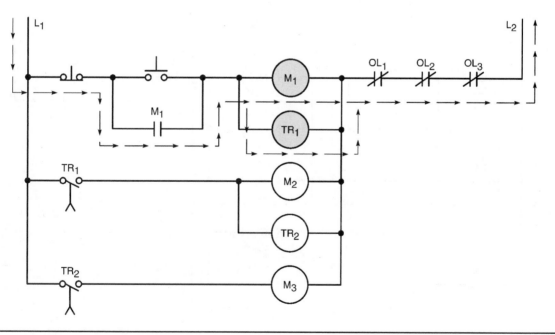

Figure 6–10B   Motor starter M_1 and time relay TR_1 turn on.

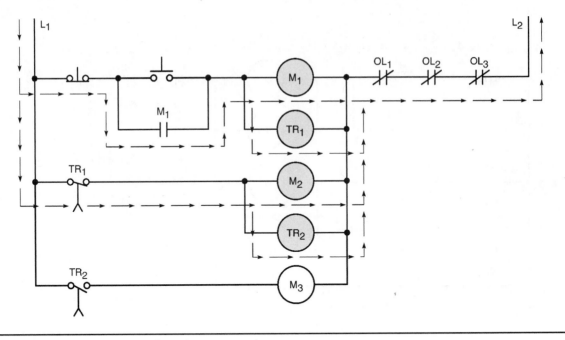

Figure 6–10C Motor #2 and TR_2 have energized.

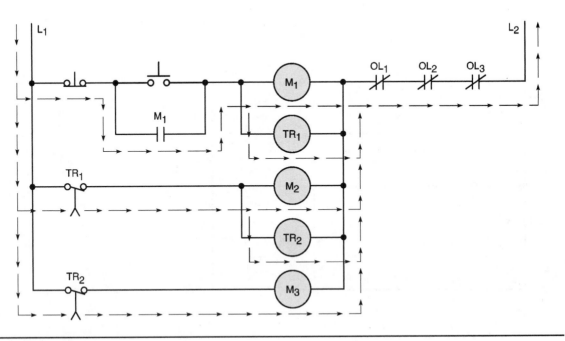

Figure 6–10D Motor #3 has energized.

When contact TR_1 opens, motor starter M_2 deenergizes, which stops motor #2, and coil TR_2 deenergizes. Since TR_2 is an on-delay relay, contact TR_2 opens immediately. This breaks the circuit to motor starter M_3. When motor starter M_3 deenergizes, motor #3 stops. Although the explanation of what happens when the stop button is pressed is quite lengthy, the action of the relays is almost instantaneous. If one of the overload contacts opens while the circuit is energized, the effect is the same as pressing the stop button. After the circuit stops, all contacts return to their normal positions and the circuit is the same as the original circuit shown in Figure 6–10A.

Solid-State Relays

The *solid-state relay* is a device that has become increasingly popular for switching applications. The solid-state relay has no moving parts, it is resistant to shock and vibration, and it is sealed against dirt and moisture. The greatest advantage of the solid-state relay, however, is the fact that the control input voltage is isolated from the line device the relay is intended to control.

Solid-state relays can be used to control DC and AC loads. If the relay is designed to control a DC load, a power transistor is used to connect the load to the line as shown in Figure 6–11A. The relay shown in the Figure has an LED connected to the input or control voltage. When the input voltage turns the LED on, a photodetector connected to the base of the transistor turns the transistor on and connects the load to the line. This optical coupling is a very common method used with solid-state relays. The relays that use this method of coupling are said to be *opto-isolated*, which means that the load side of the relay is optically isolated from the control side of the relay. Since a light beam is used as the control medium, no voltage spikes or electric noise produced on the load side of the relay can be transmitted to the control side of the relay.

Solid-state relays intended for use as AC controllers have a triac, rather than a power transistor, connected to the load circuit (Figure 6–11B). As in Figure 6–11A, an LED is

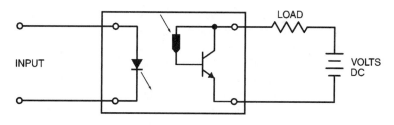

Figure 6–11A Power transistor used to control DC load.

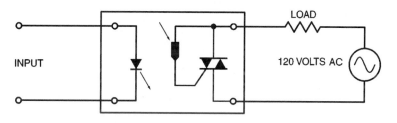

Figure 6–11B Triac used to control an AC load.

used as the control device in this example. When the photodetector *sees* the LED, it triggers the gate of the triac and connects the load to the line.

Although opto-isolation is probably the most commonly used method for the control of a solid-state relay, it is not the only method used. Some relays use a small *reed relay* to control the output. A small set of reed contacts is connected to the gate of the triac. The control circuit is connected to the coil of the reed relay. When the control voltage causes a current to flow through the coil, a magnetic field is produced around the coil of the relay. This magnetic field closes the reed contacts, causing the triac to turn on. In this type of solid-state relay, a magnetic field, rather than a light beam, is used to isolate the control circuit from the load circuit.

The control voltage for most solid-state relays ranges from about 3 to 32 volts and can be DC or AC. If a triac is used as the control device, load voltage ratings of 120 to 240 volts AC are common and current ratings can range from 5 to 25 amperes. Many solid-state relays have a feature known as *zero switching*, which means that if the relay is told to turn off when the AC voltage is in the middle of a cycle, it will continue to conduct until the AC voltage drops to a zero level and will then turn off. For example, assume that the AC voltage is at its positive peak value when the gate tells the triac to turn off. The triac will continue to conduct until the AC voltage drops to a zero level before it actually turns off. Zero switching can be a great advantage when used with some inductive loads such as transformers. The core material of a transformer can be left saturated on one end of the flux swing if power is removed from the primary winding when the AC voltage is at its positive or negative peak. This can cause inrush currents of up to 600 percent of the normal operating current when power is restored to the primary winding.

Solid-state relays are available in different case styles and power ratings. Some solid-state relays are designed to be used as time-delay relays. One of the most common uses for the solid-state relay is the *I/O* (eye-oh) track of a programmable controller, which will be covered in a later chapter.

Contactors

The *contactor*, in general, is constructed in a similar fashion to the relay. Like the relay, it is an electromechanical device. The same coil conditions exist in that a high inrush current is available when the coil of the contactor is energized. The current level drops to the holding or sealed level when the contacts are operated.

Generally, the contactor is supplied in two-, three-, or four-pole arrangements. The coil of the contactor is generally energized at 120 volts. The major difference is in the size range available with contactors. Contactors capable of carrying current in the range of 9 amperes through approximately 2250 amperes are available. For example, a size 00 contactor is rated at 9 amperes (200 to 575 volts); a size 9 contact is rated at 2250 amperes (200 to 575 volts).

One normally open auxiliary contact is generally supplied as standard on most contactors. This contact is used as a holding contact in the circuit, for example, around a normally open push-button switch. Additional normally open and normally closed auxiliary contacts can be obtained as an option. They can be supplied with the contact from the manufacturer or ordered as a separate unit and mounted in the field.

The use of the auxiliary contact, which is generally rated at 10 amperes, is shown in Figure 6–12. It can be seen that the auxiliary contact is used like the relay interlock or holding circuit. In the circuit in Figure 6–12, the light is connected across the source of electric energy (circuit lines *1* and *2*). The light in the circuit in Figure 6–12, therefore, is energized when the control *ON-OFF* selector switch is operated to the *ON* position.

CONDUCTORS CARRYING LOAD CURRENT AT LINE VOLTAGE ARE DENOTED BY HEAVY LINES.

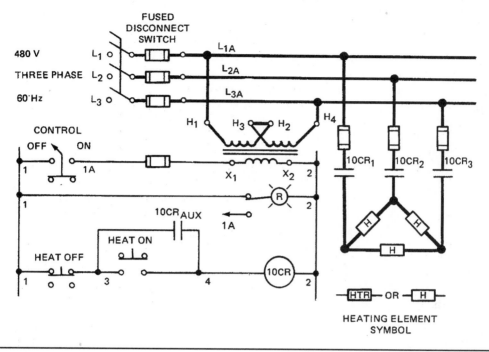

Figure 6–12 Control circuit. Courtesy Square D Company.

The balance of the sequence is as follows:

1. Operate the *HEAT ON* push-button switch.

2. Coil of contactor *10CR* is energized.

3. Power contacts *10CR$_1$*, *10CR$_2$*, *10CR$_3$* close, energizing the heating elements at line voltage.

4. Auxiliary contact *10CR$_{aux}$* closes, interlocking around the *HEAT ON* push-button switch.

5. Operate the *HEAT OFF* push-button switch, deenergizing the coil of *10CR* contactor.

   a. Contacts *10CR$_1$*, *10CR$_2$*, *10CR$_3$* open, deenergizing the heating elements.
   b. Auxiliary contact *10CR$_{aux}$* opens, opening the interlock circuit around the *HEAT ON* push-button switch.

   In addition to the conventional electromechanical contactor is the *mercury-to-metal contactor.* They are available up through 100-ampere capacity. The load terminals are isolated from each other by the glass in the hermetic seal. The plunger assembly, which includes the ceramic insulator, the magnetic sleeves, and related parts, floats on the mercury pool. When the coil is powered, creating a magnetic field, the plunger assembly is pulled down into the mercury pool, which in turn is displaced and moved up to make contact with the electrode. This closes the circuit between the top and bottom load terminal, which is connected to the stainless steel can.

The basic use for the contact is for switching of power in resistance heating elements, lighting, magnetic brakes, or heavy industrial solenoids. Contactors can also be used to switch motors if separate overload protection is supplied.

Trouble-shooting

See Troubleshooting Chart, Appendix G.

CHAPTER 7

Switches and Sensors

Drum Switches

A *drum switch* is a manually operated multiposition, multipole switch that carries a horse-power rating and is used for applications such as manual reversing of single- or three-phase motors. Drum switches are available in several sizes and can be *momentary contact* or *maintained contact*. Separate overload protection via manual or magnetic starters must usually be provided since drum switches do not include this function.

Snap Switches

Snap switches for motor control are enclosed, precision switches that require low operating forces and have a high repeat accuracy. They are used as interlocks, as the switch mechanism for control devices such as precision limit switches, and as pressure switches. They are also available with integral operators for use as compact limit switches and door-operated interlocks.

The following are examples of snap-switch operation:

1. *Single-Pole–Single-Throw* (SPST): Single-pole–single-throw switches make one contact with one throw of the switch.
2. *Single-Pole–Double-Throw* (SPDT): Single-pole–double-throw switches make one contact between two points with any of two throws of the switch.
3. *Double-Pole–Single-Throw* (DPST): Double-pole–single-throw switches make two contacts between points with one throw of the switch.
4. *Double-Pole–Double-Throw* (DPDT): Double-pole–double-throw switches make two contacts between two points with any of two throws of the switch.

See Appendix D for diagrams of SPST, SPDT, DPDT, and DPST switches.

Foot Switches

A *foot switch* is a control device operated by a foot pedal used where the process or machine requires that the operator have both hands free. Foot switches have momentary contacts but are available with latches that enable them to be used as maintained-contact devices. The foot switch generally comes enclosed in a guard to protect against unintentional operation.

Float Switches

When a pump motor must be started and stopped according to changes in water (or other liquid) level in a tank or sump, a *float switch* is used. This is a control device whose contacts are controlled by movement of a rod or chain and counterweight fitted with a float. For closed-tank application, the movement of a float arm is transmitted through a bellows seal to the contact mechanism.

Push-Button Switches

Push-button switches generally consist of two parts: the contact unit and the operator. This allows for many combinations that cover almost every application required.

Contact Units

Contact units can be obtained in blocks that contain one normally open and one normally closed contact. (Normal can be defined as not being acted upon by an external force.) Multiples of these blocks can be assembled to obtain up to four normally open and four normally closed contacts. With some switches that are now available, more units can be added. The contact rating for a typical heavy-duty, oil-tight unit is as follows:

> AC volts: 110 to 125
> Amperes normal: 6.0
> Amperes inrush: 60.0

In some cases, the contact block is base-mounted in the push-button enclosure. Thus, the units can be prewired before the cover with the operators is put in place. This method also eliminates the need for cabling conductors to the cover.

The alternate method is panel mounting. The base of the operator, with the contact block attached, is mounted through an opening in the panel. It is then secured in place by a threaded ring installed from the front of the panel. The ring is part of the operator assembly. This arrangement has the advantage of providing a space for a terminal block installation in the base of the enclosure. Thus, all connecting circuits can be terminated at an easy checkpoint. There are slight differences in the way manufacturers machine mounting holes. Therefore, modifications may be required when substituting one unit for another.

Operators

Operators are available to suit almost any application. The following are some types of operators:

- Recessed button
- Mushroom head
- Time delay
- Illuminated push-pull
- Keylock

Another operator of a special class is the *maintained-contact attachment.* Operation of one unit operator will operate the attached contacts. The contacts remain in an operated condition until the second operator is operated.

Push-Button Station

A *push-button station* is a device that provides control of a motor through a motor starter by pressing a button that opens or closes contacts. It is possible to control a motor from as many places as there are stations through the same magnetic controller. This can be done by using more than one push-button station.

Two sets of momentary contacts are usually provided with push buttons so that when the button is pressed, one set of contacts is opened and the other set is closed. Push-button stations are made for two types of service: standard-duty stations for normal applications safely passing coil currents of motor starters up to size 4, and heavy-duty stations, when the push buttons are to be used frequently and subjected to hard or rough usage. Heavy-duty push-button stations have high contact ratings.

The push-button station enclosure containing the contacts is usually made of molded plastic or sheet metal. Some double-break contacts are made of copper. However, in most push buttons, silver-to-silver contact surfaces are provided for better electric conductivity and longer life.

Since control push buttons are subject to high momentary voltages caused by the inductive effect of the coils to which they are connected, good clearance between the contacts and insulation to ground and operator is provided. The push-button station may be mounted adjacent to the controller or at a distance from it. The amount of current broken by a push button is usually small. As a result, operation of the controller is hardly affected by the length of the wires leading from the controller to a remote push-button station.

Push buttons can be used to control any or all of the many operating conditions of a motor, such as start, stop, forward, reverse, fast, and slow. Push buttons also may be used as remote stop buttons with manual controllers equipped with potential trip or low-voltage protection.

Control Station

A *control station* may contain one or more push buttons, selector switches, and pilot lights in the same enclosure. Push buttons may be of the momentary- or maintained-contact type. Selector switches are usually maintained contact or can be spring return to give momentary-contact operation. Standard-duty stations will handle the coil currents of contactors up to size 4. Heavy-duty stations have higher contact ratings and provide greater flexibility through a wide variety of operations and interchangeability of components.

The Push-Pull Operator

Another type of push-button control is the *push-pull operator*. This control contains two sets of contacts in one unit. One set of contacts is operated by pulling outward on the button, and the other set is operated by pushing the button. The head of the push-pull operator is made in such a manner as to permit a machine operator to pull outward on the button by placing two fingers behind it. The head of the control can be pushed to operate the other set of contacts.

There are two types of push-pull operators. The type used is determined by the requirements of the circuit. One type of control contains two normally open momentary contacts. Figure 7–1A shows a schematic drawing of this control. In its normal position, neither movable contact *A* nor *B* connects with either of the stationary contacts. In Figure 7–1B, the control has been pulled outward. This causes movable contact *A* to connect with two of the stationary contacts. Movable contact *B* does not connect with its stationary contacts.

If the push-pull operator is pressed, movable contact *B* connects with a set of stationary contacts as shown in Figure 7–1C. Movable contact *A* does not connect with its stationary contacts.

The second type of push-pull operator has one normally open contact and one normally closed contact. Figure 7–1D shows a schematic drawing of this type of control. When this type of push-pull operator is in its normal position, movable contact *A* is open and movable contact *B* is closed. When the button is pulled outward, contact *A* connects with its stationary contacts and contact *B* maintains connection with its stationary contacts. Figure 7–1E illustrates this condition. When the button is released, the movable contacts return to their normal position. When the push button is pressed, contact *B* breaks the connection with its stationary contact. Figure 7–1F shows this condition.

This type of push-pull operator can be used as a start-stop motor-control station. The advantage of this type of control is that it requires the space of only one push-button control element instead of two. Also, since the control must be pulled outward to start the motor and pressed inward to stop the motor, the possibility of accidentally pushing the wrong button is eliminated.

A start-stop control circuit using the push-pull operator is shown in Figure 7–1G. When the push-pull operator is pulled outward, contact *A* completes a circuit to motor-starter coil *M*. Normally open contact *M*, connected parallel to contact *A*, closes to maintain the circuit to coil *M* when the button is released. When the push-pull operator is pressed, movable contact *B* breaks the circuit to motor-starter coil *M* and the circuit deenergizes.

Selector Switches

Selector switches can be obtained with up to four positions. They can be the maintained-contact type, or the three-position switch can be arranged for spring return from the right, from the left, or from both right and left. Up to eight contacts are available per device.

The operators are available in standard-knob, knob-lever, or wing-lever types. A cylinder lock can be used and the switch locked in any one position or all positions. The

Figure 7–1A Push-pull operator with two normally open contacts.

Figure 7–1B Movable contact *A* connects with the stationary contact.

Figure 7–1C Movable contact *B* connects with the stationary contact.

Figure 7–1D Movable contact *A* is normally open, and movable contact *B* is normally closed.

Figure 7–1E Both movable contacts connect with their stationary contacts.

Figure 7–1F Both movable contacts break connection with their stationary contacts.

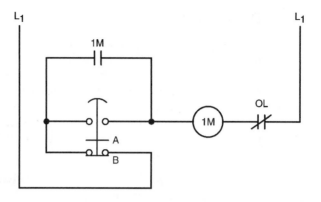

Figure 7–1G Push-pull operator used as a start-stop motor control.

arrangement for opening or closing contacts in any one position or more depends on a cam in the operator.

In the more simple form, a selector switch with two contacts arranged for two positions may be called a *double-pole–double-throw selector switch.* Similarly, a selector switch with two contacts arranged for three positions may be called a *double-pole–double-throw-with-neutral selector switch.* As the number of positions increases to four and the number of contacts (poles) increases to eight, the manufacturer's reference to a specific operator is generally coded through a symbol chart or function table. Such charts or tables display which contacts are closed or open in the different positions of the selector-switch operator.

Selector switches usually have *maintained-contact positions,* with three and sometimes two selector positions. Selector-switch positions are made by turning the operator knob—not pushing it.

Pressure Switches

Any industrial application that has a pressure-sending requirement can use a *pressure switch.* A large variety of pressure switches are available to cover the wide range of control requirements for pneumatic or hydraulic machines such as welding equipment, machine tools, high-pressure lubricating systems, and motor-driven pumps and air compressors.

Limit Switches/ Position Switches/ Sensors

A *limit switch* or *position switch* is a control device that converts mechanical motion into an electric-control signal. Its main function is to limit movement, usually by opening a control circuit when the limit of travel is reached. Limit switches may be momentary-contact or maintained-contact types. Among other applications, limit switches can be used to start, stop, reverse, slow down, speed up, or recycle machine operations.
See Chapter 2 for the application of limit switches.

Traditionally, a limit switch was a basic snap switch in an enclosure with an actuator and was used to detect the presence of cams, boxes, parts of machines, cars, elevators, and so on. A snap switch is excellent for cyclic applications since the rapid-contact transfer caused by the internal toggle springs reduces arcing and prolongs contact life. However, the toggle springs cannot exert sufficient force to break even a light contact weld caused by an overcurrent. Therefore, where the operation of the contacts is critical, as in an emergency end-of-travel application, direct-acting normally closed contacts that are directly forced open by the actuator should be used.

Definition of Terms

In the United States, the term *limit switch* has been used to refer to these devices whether they had snap-acting or direct-acting contact mechanisms. However, the *European Standards* define a limit switch as having direct-acting or positive-break contacts, and *position switch* is the general term for any switch that detects the position or presence of a cam, box, or other component.

There are many terms common to position or limit switches that are not used with other control devices. Before proceeding further, definitions of the commonly used terms should be understood.

> *Position or limit switch*   A device that converts a mechanical motion into an electric control signal (see Figure 7–2).
> *Actuator*   Mechanism of a position switch that operates the contacts, i.e., lever arm, plunger, wobble stick (see Figure 7–2).
> *Cam*   Machine part or component that applies force to the switch actuator causing it to move as intended. Also known as *dog* (see Figure 7–2).
> *Cam track dimension*   Distance from the switch-mounting surface to some point on the roller or actuator (see Figure 7–3).

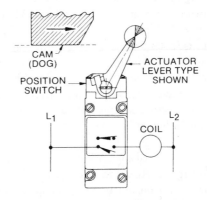

Figure 7–2   Limit switch contacts. Courtesy Square D Company.

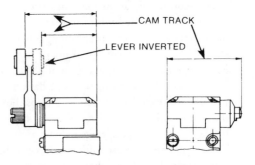

Figure 7–3   Cam track. Courtesy Square D Company.

Differential Distance the position switch actuator moves from the trip point back to the reset point of the contacts.

Direct-acting contacts Contacts are moved directly by the operating shaft. They are slow-make–slow-break contacts and have a shorter life than snap-action contacts due to longer arcing times. In general, these should only be used where movement of the actuator must break welded contacts, as in a crane safety limit switch.

Maintained contact Contacts remain in the tripped position until the return travel of the cam moves the switch actuator back and resets the contacts.

Neutral-position limit switch Lever-arm-type switch with two sets of contacts, one of which operates when the shaft is rotated clockwise and the other that operates when the shaft is rotated counterclockwise.

Operating force Force required to move position switch actuator to trip the contacts.

Overtravel Distance the position switch actuator moves beyond the trip point.

Pole Refers to the number of movable contacts in a switching mechanism. A single-pole device may be one normally open, one normally closed, or one normally open and one normally closed, but a single set of movable contacts is used to bridge those stationary contacts. A double- or two-pole switch has two movable contacts.

Positive-break contacts Contacts with a special mechanism to ensure opening. Can be snap-acting, positive-break or direct-acting, slow-make–slow-break type. The direct-acting type is not recommended for high-cycle applications due to shorter life.

Pretravel Distance the position switch actuator must move to trip the contacts.

Reed switch Mechanism that consists of a set of contacts hermetically sealed in a glass envelope and actuated by a magnet attached to the operator. This sealed construction keeps contaminants out of the contact area, making the reed switch ideal for low-voltage, low-current circuits such as programmable controllers.

Reset point The position of the actuator at which the contacts return to the normal position.

Slow-make–slow-break contacts The speed of contact transfer is directly dependent on the speed of the operating shaft. They have a shorter life due to the longer arcing times and should not be used unless required by the application.

Snap-action contacts Contacts rapidly move to open or closed position relatively independent of cam speed. Snap-acting contacts have longer contact life than slow-make–slow-break contacts due to shorter arcing time and should be used where slow-moving cams are encountered or where good repeat accuracy is required.

Solid-state switching Switching accomplished through the use of solid-state electronic components. Since there are no moving mechanical parts, it offers higher reliability and longer life than mechanical contacts. However, ratings are generally lower than those of hard contacts.

Spring return Contacts return to their original position when the actuating force is removed.

Two-stage position switch Device with two sets of contacts, one of which operates before the other. Device can be wired for either overlapping contacts or sequencing applications.

Applications

Position switches are pilot-control devices and are commonly referred to as sensors or input devices. They provide control-system-logic elements with information about the position of moving objects. Logic elements such as relays, starters, and programmable controllers can then control the actions of a machine. For example, a machine cam coming in contact with the actuator of a mechanical position switch might initiate a control sequence to stop, slow down, reverse, or recycle the machine movement. In another application, an object passing through the sensing field of a proximity or photoelectric sensor may trigger a counting, sorting, or packaging sequence. Typical applications include:

> Machine tools
> Presses
> Conveyor systems
> Elevators
> Packaging machines
> Automatic doors
> Hoists

Types of Position Switches

There are two general categories or classifications of position switches on the market: mechanical and proximity switches. *Mechanical switches* are operated when a cam physically moves the actuator, operating a set of contacts. *Proximity switches* (also called *no-touch switches*) do not come in contact with the cam and are usually solid-state devices.

Proximity Switches   *Proximity switches* are chosen where the objects to be detected cannot be touched, or where long-distance, high-speed, or high-cycle sensing is desired. They include inductive, capacitive, magnetic, ultrasonic, and photoelectric types. They may also be subdivided into interrupted-beam or reflective types. Inductive and photoelectric switches are the most common, and a brief description of their operating principles follows. For more complete information, see the referenced publications.

Inductive   *Inductive* proximity switches are all-metal-sensing, solid-state devices and operate on the following principle: An oscillator circuit is connected to a coil, producing a sensing field. When a metal object (target) passes into the field, eddy currents are induced in the target, causing a change in oscillator voltage. This voltage change is detected, and the amplifier produces a change in the switch output. It is not necessary for the object to come in direct contact with the switch. Available styles include cylindrical, rectangular, and modular turret head.

Photoelectric   A *photoelectric* switch is a noncontact sensor employing transmitted or reflected light as the sensing medium and detects all nontransparent materials. It consists of a light source (emitter) and a detector (receiver), which are contained in the same enclosure or separate enclosures. Depending on the mode of operation, it will give an output when the light path between the source and detector is completed (light-operate mode) or blocked (dark-operate mode).

Mechanical Switches   In *mechanical switches,* a rotary arm or push rod on the switch housing is mechanically connected to the electric switching element inside. The moving object comes into direct contact with the switch. Types include lever-operated, push-rod, wobble-stick, and rotary-cam. Mechanical position switches, because of low cost and versatility, are the most common type used in industry today.

Micro Switches   Another type of limit switch often used in different types of control circuits is the micro limit switch or *microswitch*. Microswitches are much smaller in size than the limit switch, which permits them to be used in small spaces that would never be accessible to the larger device. Another characteristic of the microswitch is that the actuating plunger requires only a small amount of travel to cause the contacts to change position. The microswitch has an activating plunger located at the top of the switch. This switch requires that the plunger be depressed approximately 0.015 inch or 0.38 millimeter. Switching the contact position with this small amount of movement is accomplished by spring-loading the contacts. A small amount of movement against the spring will cause the movable contact to snap from one position to another.

Electrical ratings for the contacts of the basic microswitch are generally in the range of 250 volts AC and 10 to 15 amperes depending on the type of switch. The basic microswitch can be obtained with a variety of activating arms.

Subminiature Microswitches   The *subminiature microswitch* employs a spring contact arrangement similar to that of the basic microswitch. The subminiature switches are approximately one-half to one-fourth the size of the basic switch, depending on the model. Due to their reduced size, the contact rating of subminiature switches ranges from about one ampere to about seven amperes, depending on the switch type.

Common Mechanical Switches   Lever-arm type: A passing cam operates the *lever arm,* which in turn rotates a shaft and operates the switch contacts. The switch may be spring return momentary contact or maintained contact.

Lever arms may be roller or rod type and come in various sizes. Lever-arm length is defined as the dimension from the center of the shaft to the center of the roller. The lever type should be considered first for any application.

Plunger types: *Plunger types* have a rod extending out of the head of the switch, which, when pushed, operates the contacts. The push-rod type must be operated by a cam moving in line with the plunger axis. The roller-plunger type can be operated in line with the axis or can be operated from the side with the cam leading edge generally restricted to 30 degrees or less.

Plunger types are used where small, controlled movements are to be detected or where lever types will not fit in the space allowed. Total travel is generally limited to one-fourth inch.

These types have an extension that will trip the contacts when operated from any direction. Wobble-stick position switches are generally furnished with a wire or rod extension. Cat-whisker types have nylon-covered wire extensions and have lower operating-force requirements.

These types of devices are generally used on conveyor systems to count or detect objects moving on the conveyor. Wobble sticks are used for detecting large, heavy objects, while cat whiskers are used to detect lightweight objects. These types can also be used as safety devices where a machine operator can deflect the switch by hand from any direction to shut off the machine in an emergency.

Common Mechanical Position Switch Enclosures

Surface-Mounting Type   If at all possible, a *surface-mounting switch* should be used, as this is the most common and simplest type of mounting. Generally, only two holes need be drilled and tapped for switch mounting.

Plug-In Position Switches   Downtime in an automated factory is very expensive. When a switch is damaged, it must be replaced quickly. With non-plug-in switches, the damaged unit must be removed and unwired and a new unit installed and rewired—a time-consuming operation. With a *plug-in switch,* the receptacle is mounted and wired and becomes a permanent part of the machine. When a plug-in switch is damaged, the plug-in part can be removed and a new unit plugged in quickly, reducing downtime to a minimum.

Flush-Mounting Type   *Flush-mounting switches* are used where a machine cavity is provided and the standard position switch box is not required. A gasket on the flush plate prevents entrance of contaminants. Wiring is brought in through the machine cavity.

Manifold-Mounting Type   Occasionally, it may be more convenient to run wiring in a wiring trough, panel, or raceway and bring the wires into the switch through the underside instead of through the standard conduit hole. *Manifold-mounting switches* are provided with a gasket and hole in the mounting surface and the conduit hole plugged. The gasket prevents leakage into the manifold hole.

Hazardous-Location Type   Explosive dusts or gases may be present in the atmosphere, and special devices are required in these locations. The NEMA 4, 6P, 7, and 9 explosionproof enclosure meets *NEC*® requirements for Division 1 and Division 2: Class I, Groups B, C, and D gases; and Class II, Groups E and G. The type C, reed contact switches are approved for use in Division 2 hazardous locations.

Position Switch Selection

Selecting the proper limit switch for a given application is relatively simple. Use a selection guide provided by the manufacturer. The following paragraphs offer some information to aid in selection.

Heavy-Duty Precision Oiltight Switch

The *heavy-duty precision oiltight switch* provides superior electric and mechanical life along with easy installation and wiring. It is available in a variety of head and body styles, including an explosionproof version that is also watertight and submersible. Also available are standard, logic-reed, and power-reed contacts, as well as many other options. The heavy-duty oiltight switch is useful if the load exceeds the contact ratings, if a required operating sequence is not available, or if high trip and reset forces are required. This type should be the first choice for all applications.

Miniature, Enclosed-Reed Switch

The *miniature, enclosed-reed switch* is a small, inexpensive, die-cast zinc switch utilizing a hermetically sealed reed for the contact mechanism. Prewiring and potting combined with the sealed reed make this switch a good choice where contact reliability and environmental immunity are required along with small size and low cost.

Heavy-Duty Oiltight-Type Foundry Switch

The *heavy-duty oiltight-type foundry switch* is for use in foundries or mills where hot, falling sand or similar foreign material could build up and cause jamming of the arm.

Precision Oiltight Switch

A *precision oil-tight switch* is suggested especially for applications requiring a lever-type switch for use in extremely low ambients or for micrometer adjustment on a plunger-type switch.

Inductive Proximity Switch

The *inductive proximity switch* is for use on machines or conveyors where it is necessary to detect the presence or position of metal parts without physical contact. It is particularly well suited for very small parts, odd-shaped parts, high-speed applications, and other applications where conventional position switches will not perform satisfactorily. When selecting a proximity switch, there are many factors to consider. Mounting, temperature range, shielding, sensing distance, and load requirements make selecting the proper switch for a particular application more involved. Covering these factors is beyond the scope of this publication. See manufacturer's specifications.

Photoelectric Switch

The *photoelectric switch* is used on machines, conveyors, or other applications where longer-range, no-touch sensing is required. It will detect almost any material and can be used for box or can counting, jam detection, people detection, and so on. A variety of detection modes and enclosures are available. Selection of photoelectric switches involves choosing the proper sensing mode and range for the application, selecting the appropriate type of enclosure, determining excess gain required, and considering the operating environment.

Lever-Arm Switch

Standard 10-degree Pretravel Lever-Arm Switches   These switches will handle about 90 percent of all applications.

Standard Clockwise (CW) and Counter-Clockwise (CCW) Version   This switch will handle most applications with no conversion necessary. Where CW only or CCW only is required, the switch can be easily converted by removing the turret head and rotating the guide until the arrowpoints to the appropriate operating-mode symbol.

Plug-in Standard Box   The trend is toward plug-in switches because of easy replacement. Also, the standard box is the same size for one or two poles.
The foregoing selection leads you to:

> Single-pole–double-throw
> Double-pole–double-throw
> Neutral position
> Two-stage

This information must be given when designing this application. Each manufacturer has model numbers for this application.

Specialty Lever-Arm Switches

Low-Differential Type   A *low-differential-type switch* is generally required where the differential must be small and should not be selected for the 5-degree pretravel feature except in the case of neutral-position application. The desired trip point can usually be obtained by adjusting the lever arm and/or cam. For the specific recommended type, see manufacturer's specifications.

Light-Operating, Torque Type   A *light-operating torque-type switch* is used where the operating torque of the standard type is too high. For the recommended type, see manufacturer's specifications.

Gravity-Operated Type   The *gravity-operated-type switch* is used where extremely light operating torque is required. The lever arm is returned by gravity. For the recommended type, see manufacturer's specifications.

Maintained-Contact Type   A *maintained-contact-type switch* is used where a memory device is required to *remember* that a cam has passed. For the recommended type, see manufacturer's specifications.

Other   If a space problem exists, select the *compact box;* see manufacturer's specifications. If other enclosure types are required, see manufacturer's specifications.

Types of Lever Arms Available for All Mechanical Switches

1. Standard lever arms for position switches:

   Lever arm, 1 ½ inches long with a ¾-inch diameter, ¼-inch wide roller.

   Maintained contact: For resetting by the same cam, select a forked lever arm. See manufacturer's specifications.

2. Position switch, lever arm, 1 ½ inches long with a ¾-inch diameter, ¼-inch wide roller. See manufacturer's specifications.

3. Lever arm, 1 ⅜ inches long with a ⅝-inch diameter, ¼-inch wide roller. See manufacturer's specifications.

Plunger-Type Switches

Plunger-type switches are used where short, controlled machine movements are present and where space or mounting does not permit a lever-arm switch. Recommended types of plunger-type switches are:

Top-roller plunger—see manufacturer's specifications.
Side-roller plunger—see manufacturer's specifications.
Top push-rod plunger—see manufacturer's specifications.
Side push-rod plunger—see manufacturer's specifications.

If space or precise control of the trip point is an issue, see manufacturer's specifications to determine the proper size needed.

Wobble-Stick and Cat-Whisker Switches

Wobble-stick and *cat-whisker switches* are suitable for application on conveyors to detect or count parts or as hand-operated safety devices. Wobble-stick and cat-whisker switches can be operated from any direction. Cat-whisker switches are used to detect very lightweight parts. Recommended types are:

Wobble, Delrin extension—see manufacturer's specifications.
Wobble, wire extension—see manufacturer's specifications.
Cat Whisker—see manufacturer's specifications.

Ratings

In order to apply a control device correctly, the ratings of the switch must be compared to the load. Three types of ratings are generally given for any switch:

1. *Resistive or noninductive ratings*—This rating indicates the resistive load (e.g., lighting or heating) that the contacts can make and break. Resistive ratings are generally based on a 75 percent power factor for AC.

2. *Inductive (pilot duty) ratings*—Most control applications today involve highly inductive loads such as starters, contact and relay coils, solenoids, and clutches. These types of loads have a high inrush of current upon energization. As the load is sealed in, the current drops significantly. Both the inrush and sealed values must be known and compared to the make-and-break ratings of the switch to assure compatibility. These ratings are based on a 35 percent power factor for AC.

3. *Continuous Rating*—Continuous rating indicates the load that the contacts can carry continuously (thermal rating) without making or breaking the circuit. The *heavy-duty* rating is more than sufficient for most applications. However, the larger, solenoid-coil loads may require the *extra-heavy-duty* ratings.

   Example: Load is a NEMA size 1 starter coil with the following characteristics:
   Coil-120 volts AC 60 Hertz
   Inrush 245 volt-amperes
   Sealed 27 volt-amperes
   Inrush amperes are 245/120 or 2.04 amperes.
   Sealed amperes are 27/120 or .225 amperes.

Limit switches can easily make 2.04 amperes and break .225 amperes at 120 volts.

For maximum position switch life, the force applied to the lever arm by a cam should be perpendicular to the lever arm. This means that the cam angle and the lever-arm angle should be the same. The force applied should also be perpendicular to the shaft axis about which the lever rotates.

Recommend-ations	## Non-Overriding Cams

The arrangement shown in Figure 7–4 is satisfactory only at cam speeds below 50 feet per minute. At higher speeds, the impact due to high lever acceleration causes excess roller bounce.

NON-OVERRIDING CAMS

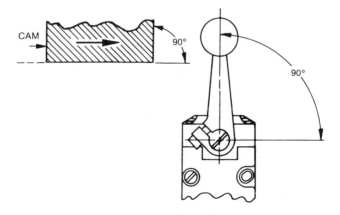

Figure 7–4   Non-overriding cam. Courtesy Square D Company.

A good recommended cam angle and lever-arm angle at moderate cam speeds (up to 90 feet per minute) is 45 degrees (Figure 7–5). Here, lever acceleration is less, and deceleration is also less at the lower cam edge.

Overriding Cams

The cam trailing edge on overriding cams must also be considered for maximum switch life (see Figure 7–6). Lever-arm snapback causes shock loads, which reduce switch life. Also, with reversing cams, the trailing edge becomes a leading edge on the return stroke.

The overtravel of the position switch should not be exceeded, but a minimum of 5 degrees to 15 degrees travel past the trip point is recommended. Additional travel should only be used for setup and emergencies. Cam design procedures for position switches with other than lever-arm actuators vary from switch type to switch type.

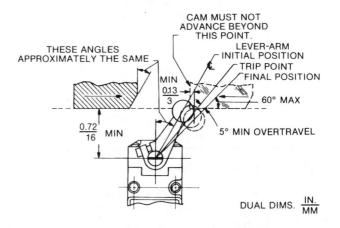

Figure 7–5   Non-overriding cam. Courtesy Square D Company.

OVERRIDING CAMS

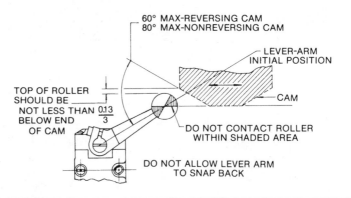

Figure 7–6   Overriding cam. Courtesy Square D Company.

Contacts

1. Make sure the electric load is within position switch contact ratings.

2. The single-pole–double-throw contacts of a snap switch used in a position switch should not be used on opposite polarities. When load M_1 is connected between the contact and line L_2 and load M_2 is connected between the other contact and line L_1, a line-to-line short can occur through the arc, which may be drawn as the contacts operate. When contacts are connected to the same polarity, this line-to-line short cannot occur. The same result can occur if different power sources are connected to the single-pole–double-throw contacts of a snap switch.

3. When applying position switches having reed contacts, it is suggested that some form of transient protection be used. This will protect the small contacts from damaging surges and will increase contact life.

Coolant

1. When possible, avoid mounting position switches where they will be constantly exposed to coolant, chips, and so on. Although designed for such applications, switches obviously will last longer when not exposed to these contaminants.

2. Make sure cover screws are tightened to assure a good, oiltight seal.

3. When possible, avoid use of fire-resistant coolants of the phosphate-ester type. Equipment exposed to these coolants requires special seals and gaskets.

Installation of Conduit

Position-switch leakage is very often traced to the conduit system. Coolant or condensation in the conduit line can enter the switch through the conduit entry. Oiltightness is dependent upon the condition of the conduit connection and seal. Recommendations for installing conduit to position switches are as follows:

1. To ensure an oiltight seal, use thread sealant and a conduit seal or a sealing bushing around the conduit fitting. If this is not done, there is a good possibility that the fitting will leak.

2. Position switches should be installed with the conduit end down whenever possible.

3. If condensation or moisture is present inside the conduit, a conduit seal can be inserted in the conduit entry. The conduit fitting can then be connected in the normal manner. Thread sealant and a sealing bushing must still be used.

4. Often a junction box fills with coolant and/or condensation, which backs up into the position switch through the conduit. A simple solution is to drill a hole in the bottom of the junction box to allow the liquid to drain out.

5. If conduit leakage is severe, prewired and potted position switches should be used. The switches are prewired with either individual wires or multiconductor STOWA cord, and the receptacle is completely sealed with a potting material.

6. The limit (position) switch is available with a prewired male plug receptacle. The connector provides an effective oiltight seal when used with the appropriate female connector cord.

Sensors

Certain applications require the use of *no-touch sensing*. Many varieties of *sensors* are available, most falling into one of two categories: proximity and photoelectric. *Proximity sensors* detect the disturbance of an electric field generated by the device. *Photoelectric sensors* detect the breaking of a beam of light.

The control of pumps, air compressors, welding machines, lube systems, and machine tools requires control devices that respond to the pressure of a medium such as water, air, or oil. The control device that does this is a *pressure switch*.

The pressure switch has a set of contacts that are operated by the movement of a piston, bellows, or diaphragm against a set of springs. The spring pressure determines the pressures at which the switch closes and opens its contacts.

Pressure Sensors

Pressure sensors are designed to produce an output voltage or current that is dependent on the amount of pressure being sensed. *Piezoresistive sensors* are very popular because of their small size, reliability, and accuracy. These sensors are available in ranges from zero to one psi (pound per square inch) and zero to thirty psi. The sensing element is a silicon diaphragm integrated with an integrated circuit chip. The chip contains four implanted piezoresistors connected to form a bridge circuit.

When pressure is applied to the diaphragm, the resistance of piezoresistors changes proportionally to the applied pressure, which changes the balance of the bridge. The voltage across VOM changes in proportion to the applied pressure ($V = V4 - V2$ [when referenced to V3]). Typical millivolt outputs and pressures are shown below:

> 1 psi = 44 millivolts
> 5 psi = 115 millivolts
> 15 psi = 225 millivolts
> 30 psi = 315 millivolts

This particular sensor can be used to sense absolute, gauge, or differential pressure. Units are available that can be used to sense vacuum. Sensors of this type can be obtained to sense pressure ranges of 0 to 1, 0 to 2, 0 to 5, 0 to 15, 0 to 30, and 0 to –15 (vacuum). The sensor contains an internal operational amplifier and can provide an output voltage proportional to the pressure.

Typical supply voltage for this unit is 8 volts DC. The *regulated* voltage output for this unit is 1 to 6 volts. Assume for example that the sensor is intended to sense a pressure range of 0 to 15 psi. At zero psi, the sensor would produce an output voltage of one volt. At 15 psi, the sensor would produce an output voltage of 6 volts.

Sensors can also be obtained that have a ratiometric output. The term *ratiometric* means that the output voltage will be proportional to the supply voltage. Assume that the supply voltage increases by 50 percent to 12 volts DC. The output voltage would increase by 50 percent also. The sensor would now produce a voltage of 1.5 volts at zero psi and 9 volts at 15 psi.

Other sensors can be obtained that produce a current output of 4 to 20 milliamperes, instead of a regulated voltage output. One type of pressure-to-current sensor that can be used senses pressures as high as 250 psi. This sensor can also be used as a point detector to provide a normally open or normally closed output. Sensors that produce a proportional output current instead of voltage have fewer problems with induced noise from surrounding magnetic fields and with voltage drops due to long wire runs.

Flow Sensors*

Flow sensors are used to detect liquid flowing through a pipe or air flowing through a duct. Flow switches, however, cannot detect the amount of liquid or airflow. To detect the amount of liquid or airflow, a *transducer* must be used. A transducer is a device that converts one form of energy into another. In this case, the kinetic energy of a moving liquid or gas is converted into electric energy. Many flow sensors are designed to produce an output current of 4 to 20 milliamperes. This current can be used as the input signal to a programmable controller or as the input to a meter designed to measure the flow rate of the liquid or gas being metered.

Liquid Flow Sensors   There are several methods that can be used to measure the flow rate of a liquid in a pipe. One method uses a *turbine sensor*. The turbine sensor consists of a turbine blade that must be inserted inside the pipe containing the liquid. The moving liquid causes the turbine blade to turn. The speed at which the blade turns is proportional to the amount of flow in the pipe. The sensor's electric output is determined by the speed of the turbine blade. One disadvantage of the turbine sensor is that the turbine blade offers some resistance to the flow of the liquid.

Electromagnetic Flow Sensors   Another type of flow sensor is the *electromagnetic flow sensor*. These sensors operate on the principle of *Faraday's Law* concerning conductors moving through a magnetic field. This law states that when a conductor moves through a magnetic field, a voltage will be induced into the conductor. The amount of induced voltage is proportional to the strength of the magnetic field and the speed of the moving conductor. In the case of the electromagnetic flow sensor, the moving liquid is the conductor. As a general rule, liquids should have a minimum conductivity of about 20 micro-ohms per centimeter.

Flow rate is measured by small electrodes mounted inside the pipe of the sensor. The electrodes measure the amount of voltage induced in the liquid as it flows through the magnetic field produced by the sensor. Since the strength of the magnetic field is known, the induced voltage will be proportional to the flow rate of the liquid.

Airflow Sensors   Large volumes of *airflow* can be sensed by propdriven devices similar to the liquid flow sensor. Solid-state devices are commonly used to sense smaller amounts of air or gas flow. This device operates on the principle that air or gas flowing across a surface causes heat transfer. The sensor contains a thin-film, thermally isolated bridge with a heater and temperature sensors. The output voltage is dependent on the temperature of the sensor surface. Increased airflow through the inlet and outlet ports will cause a greater amount of heat transfer, reducing the surface temperature of the sensor.

Inductive Proximity Sensors†

Traditionally, the mechanical limit switch has been the dominant leader in position sensing devices. However, the latest control technology is demanding more solid-state sensors as control elements in automated processes. There is an increasing number of applications

---

*Adapted with permission from Herman/Aldrich *Industrial Motor Controls,* Delmar Publishers.

†Adapted with permission from Square D Company.

in which noncontact (solid-state) sensing provides a better solution than contact (mechanical) sensing. Due to the noncontact sensing and logic capabilities of solid-state devices, they can provide more reliable information in advanced automated systems.

With solid-state proximity sensors, physical contact with the object being detected is not required. These devices have the capability to sense components directly and can operate at much higher switching rates than mechanical limit switches. Since proximity sensors have no moving parts (solid state), they can offer a long and maintenance-free life. Inductive proximity sensors are being used by industries of all types as the need for greater reliability and speed becomes more apparent each day.

Definition of Terms

Axial approach   When the target to be detected approaches the sensing face head-on.

Complementary outputs   Sensors with both normally open and normally closed outputs that change state simultaneously.

Current-sinking sensor (NPN transistor)   A current-sinking sensor *sinks* current from the load to the negative terminal (–) of the DC voltage supply (Figure 7–7).

Current-sourcing sensor (PNP transistor)   A current-sourcing sensor *sources* current from the positive terminal (+) of the DC voltage supply to the load (Figure 7–8).

Differential (hysteresis)   The distance between the operating point where the target enters the sensing field (sensor energizes) to the release point where the target leaves the sensing field (sensor deenergizes).

Lateral approach   When the target to be detected approaches the sensing face from the side (slide-by).

Line-powered sensor (three or four-wire)   A sensor that draws its operating current (burden current) directly from the line. Its operating current does not flow through the load, and a minimum of three connections (three-wire) are required. A four-wire sensor has complementary outputs and requires four connections (Figure 7–9).

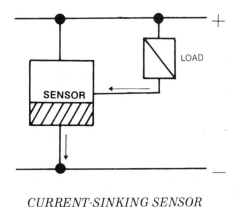

CURRENT-SINKING SENSOR

Figure 7–7   Current-sinking sensor. Courtesy Square D Company.

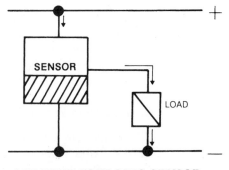

CURRENT-SOURCING SENSOR

Figure 7–8   Current-sourcing sensor. Courtesy Square D Company.

Load-powered sensor (two-wire)   A sensor that draws its operating current (leakage current) through the load. The sensor is always in series with the load, and only two connections are required (Figure 7–10).

Normally open (NO) output   The sensor closes a circuit to the load when a target is detected.

Normally closed (NC) output   The sensor opens a circuit to the load when a target is detected.

Operating distance   The distance from the sensing face to the plane of the target's path once it reaches the operating point (Figure 7–11).

Operating point   The point at which a target is sensed as it approaches the sensing field of the sensor. Also called *trip point.*

Release point   The point at which a sensor returns to its original state as the target leaves the sensing field. Also called *reset point.*

Repeat accuracy (repeatability)   The measure of variation in operating distance between successive operations under constant operating conditions. This measurement is often expressed as a maximum percentage of the *operating distance* (i.e., 5 percent).

Response time   The time delay from when a target reaches the operating point to when an output actually occurs. A target must stay within the sensing field for at least the response time, or the sensor will not change output status.

NOTE: The target must also remain within the sensing field long enough to allow the load sufficient time to respond to the output signal of the sensor.

Sensing range   The maximum operating range at which the sensor will reliably detect a standard target under conditions of nominal voltage and temperature (Figure 7–12).

Standard target   An object used to determine sensing range. This is normally a square, mild steel plate one millimeter thick, with the length of each side equal to the diameter of the sensing face or three times the nominal sensing distance of the sensor.

Target   The object to be detected.

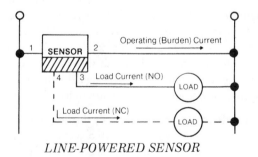

LINE-POWERED SENSOR

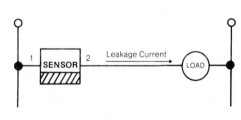

LOAD-POWERED SENSOR

Figure 7–9   Line-powered sensor. Courtesy Square D Company.

Figure 7–10   Load-powered sensor. Courtesy Square D Company.

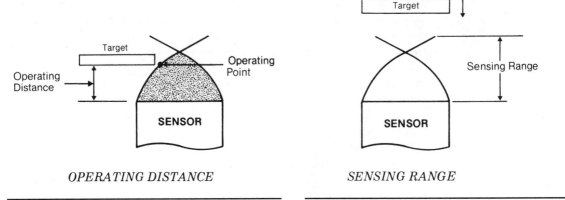

OPERATING DISTANCE

SENSING RANGE

Figure 7–11   Operating distance. Courtesy Square D Company.

Figure 7–12   Sensing range. Courtesy Square D Company.

The Inductive Proximity Sensor Defined   An *inductive proximity sensor* is a solid-state input device that can sense any metal object without physically touching it. These devices are available in many different shapes and sizes to meet the requirements of most applications. The style selected is usually determined by the ease of installation, space limitation, required sensing range and environmental conditions. Because this sensor is solid state (no moving parts), the electronics can be completely sealed (potted), making it virtually immune to shock and vibration while guarding against the entry of dust, dirt, and other contaminants.

How Does It Work?   The internal circuitry of a proximity sensor typically consists of:

- Sensing coil and oscillator circuit
- Detector circuit
- *Solid-state* output switch

Inductive proximity sensors use the *eddy-current-killed-oscillator (ECKO)* principle for detecting metal objects. The heart of the ECKO principle is an oscillator connected to a sensing coil generating a high-frequency electromagnetic sensing field that radiates from the sensing face of the sensor (see Figure 7–13). When a metal object enters the electromagnetic sensing field, eddy currents (I) are induced on the surface of the metal object (see Figure 7–13). These eddy currents reduce the amplitude of the oscillating circuit, and, when it drops to a predetermined level, the detector circuit signals the *solid-state* output switch to change state. When the metal object passes out of the sensing field, the oscillator builds back up to its normal level, allowing the output switch to return to its original state.

Where Should It Be Used?   Inductive proximity sensors are primarily used for short-range sensing (generally less than two inches) where contact with the metal object to be sensed is not desired. They are also used in applications requiring high-speed counting where mechanical limit switches would have a short life.

If any of the following conditions exist, an inductive proximity sensor should be used:

- The metal object being detected is too lightweight to be sensed mechanically.
- Fast response time and high switching frequencies are required.

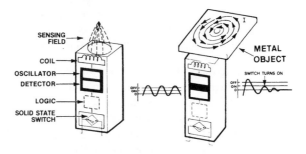

Figure 7–13 Sensing field. Courtesy Square D Company.

- Extremely long life and reliable performance are required.
- Direct interfacing with electronic control equipment (especially low-level DC) is required.

Typical applications are:

- Machine tools
- Packaging machines
- Conveyor or transfer lines
- Assembly machinery
- Robotics

Typical uses are:

- Parts counting and sorting
- Position and presence control
- Void/jam control
- Missing parts control
- Damaged tool detection
- Alignment control

What Are the Various Style Enclosures? Inductive proximity sensors are available in many different shapes and sizes. Therefore, only the most commonly used styles such as cylindrical and rectangular will be covered.

Cylindrical Style The *cylindrical style* is ideal where mounting space is limited, and its threaded barrel design is very useful for fine adjustments to the sensing point. Various diameters commonly used are: 5, 8, 12, 18 and 30 millimeters.

Rectangular Styles The rectangular turret-head styles have enclosures similar to mechanical limit switches and offer familiar limit-switch-type mounting. Unlike the cylindrical styles where the sensing face is always located at the end of the cylinder, a turret-head construction means that the sensing face can be rotated or located in one of five positions (four sides and one end location). Also, some turret-head constructions enable the head to be rotated at a 45-degree angle, thus allowing up to nine different positions.

The rectangular long-range styles have an oversized sensing face for greater sensing distance capability. These units are primarily used when standard (cylindrical and rectangular) devices cannot provide the sensing range required for an application.

Sensing-Range Considerations In general, most inductive proximity sensors worldwide are designed to meet a European standard. This standard establishes the *sensing range* and the *standard* size target for each size sensor. Proximity sensors are available with various sensing ranges—typically one millimeter to 120 millimeters. Some versions have adjustable sensitivity for varying the range. As a general rule, the longer the sensing range, the larger the sensing-face diameter and the total device.

To ensure proper operation of inductive proximity sensors, the user should understand the factors that may affect the electromagnetic sensing field. The sensing range is dependent on the following factors:

- Manufacturing tolerances
- Ambient conditions
- Target material and size
- Application factors

Manufacturing Tolerances The solid-state internal circuitry of proximity sensors is made up of several components such as resistors, capacitors, diodes, and hybrid circuits. Each component is designed to perform within a specified *tolerance,* so the combined tolerances of each component will result in a *system tolerance.* This is more commonly referred to as *manufacturing tolerances,* which cause the sensing range to vary from sensor to sensor within a production *batch.*

According to the European standard, any proximity sensor is allowed to have a sensing range of +10 percent of what is published to allow for manufacturing tolerances. Most manufacturers will list the published sensing range and a plus or minus percentage to account for manufacturing tolerances. For example, sensors with a published sensing range of 15 millimeters +/–10 percent, will detect a standard target between 13.5 millimeters and 16.5 millimeters. However, when selecting a sensor for an application, always use the lower end of the range (13.5 millimeters) for worse-case conditions.

Ambient Conditions Variations in temperature, operating voltage, and/or humidity may cause the sensing range to drift. A change in any or all three of these factors will result in an additional sensing-range variation of +/–10 degrees (typical). Therefore, the total possible sensing-range variation due to manufacturing tolerances and changes in temperature, voltage, or humidity is +/–20 degrees. So, it is recommended that the usable sensing range be not more than 80 degrees of the published range.

Example:

15 millimeters (published range) × 80 percent = 12 millimeters (usable sensing range)

Target Material and Size The sensing range of an inductive proximity sensor is based on a *standard*-size metal target. Any variation in the target size or metal composition will affect (in most cases, reduce) the usable sensing range.

Target material: With a standard-size mild steel target, maximum sensitivity can be achieved without affecting the usable sensing range. Metals other than mild steel will more than likely reduce the usable sensing range. Therefore, most manufacturers provide correction factors to allow an estimation of sensing range when the target's metal composition is something other than mild steel.

Conductive metals can be divided into two categories: ferrous and nonferrous. Ferrous metals contain iron, cobalt, or nickel and exhibit strong magnetic properties, while non-ferrous metals such as aluminum, brass, or copper do not. As a result, ferrous metals can be detected at much longer distances.

Typical correction factors follow:

Target Composition	Correction Factor Multiplier
Mild steel	1.0
Stainless steel	.925
Brass	.45
Aluminum	.40
Copper	0.35

To apply, simply multiply the correction factor multiplier by the usable sensing range (published range × 80 percent) to obtain the usable sensing range for metals other than mild steel (see Figure 7–14).

Target size: When a target enters the sensing field, eddy currents are induced on the surface, which reduces the amplitude of the oscillator. Since targets may vary in size and shape from one to the next, it is virtually impossible to develop an estimated percentage to account for the reduction in sensing range. The standard-size target for any proximity sensor is selected in order to achieve maximum sensitivity of the sensing coil. Therefore, targets larger than the standard size will not necessarily increase the sensing range due to the physical size of the sensing coil.

If the target used is smaller than the standard-size target for a particular sensor, it must pass much closer to the sensing face to sufficiently damp the oscillator. Therefore, smaller targets have the effect of reducing the sensing range. It is recommended that the standard-size target be used as a guideline when trying to optimize sensitivity, even though various size targets can be used.

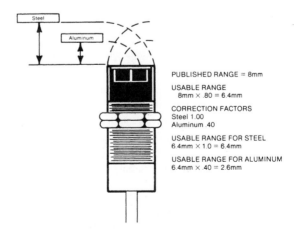

Figure 7–14 Target. Courtesy Square D Company.

Application Factors There are certain factors that should be taken into account in the application of proximity sensors. Proper operation is dependent on the following:

- *Surrounding metal:* When surrounding metal is present in an application, proper mounting clearances must be maintained. If not, the differential may be significantly affected. For example, once the target has cleared the sensing field of the device, the output may not release (return to its original state). Therefore, a metal-free zone must exist around the sensing face of the unit. The dimensions of this metal-free zone will vary depending on whether the device is shielded or unshielded.

  A *shielded* sensor head construction directs the electromagnetic field forward, restricting it from radiating out the side and detecting surrounding metal. Therefore, these sensors can be flush mounted (embedded) in metal without affecting the sensing range. However, a metal-free zone is still recommended to ensure proper operation.

  Unshielded sensors have the advantage of sensing a metal target at a greater distance because their sensing field is allowed to radiate out the side of the sensing face. These devices must be mounted above any metal framework due to their side sensitivity, which may detect surrounding metal. A metal-free zone is strongly recommended around the sensing face to eliminate false operation.

- *Adjacent proximity sensors:* When inductive proximity sensors are mounted too close to each other, their electromagnetic fields may interact. This is often referred to as *mutual interference* and may cause some sensors to operate erratically. With identical sensors (i.e., two cylindrical 18-millimeter shielded devices with the same type number), mutual interference is more likely to occur because their oscillation frequencies are typically the same. As a result, a minimum spacing is required when identical sensors are mounted adjacent to each other. On the other hand, dissimilar sensors (i.e., an 18-millimeter shielded and an 18-millimeter unshielded device) are less affected by mutual interference because their oscillation frequencies are typically different. However, if mounting space is not critical in an application, then it is still recommended that some separation be maintained to ensure proper operation.

  If for some reason it is necessary to mount sensors closer than what is normally recommended, an *alternate-frequency* device should be used. In these devices, their sensing field oscillates at a different frequency than any other *standard* sensors, which drastically reduces the possibility of mutual interference.

- *Flying metal chips:* In certain applications (milling, drilling, etc.), excessive metal chips may affect device operation, depending on the size and metal composition. In these types of applications, it is recommended that the sensor be mounted with the sensing face looking down somewhat to prevent metal chips from accumulating on it. Generally, a metal chip does not have sufficient surface area to cause a sensor to give an incorrect detection signal, but several chips on the sensing face may cause the sensor to not release (return to its original state).

Understanding a Field Map Unlike the lever arm of a mechanical limit switch, the actuator (lever arm) for an inductive proximity sensor is an invisible electromagnetic field. Therefore, a *field map* is developed to give a pictorial representation of the *effective* sensing area based on a standard target. When a sensor is purchased, a field map is normally

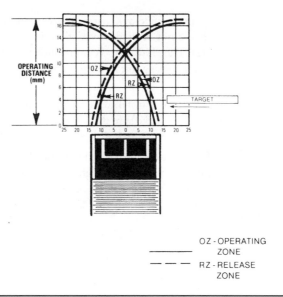

OZ - OPERATING
         ZONE

RZ - RELEASE
        ZONE

Figure 7–15 Field map. Courtesy Square D Company.

included in the instruction sheet. It is helpful in illustrating where a mild steel target must be to actuate the sensor.

A field map separates the sensing area into two parts: an *operating zone* and a *release zone*. In Figure 7–15, when the leading edge of the standard target crosses the boundary of the operating zone or operating point, the output will conduct and change state. The target in this example will actuate the sensor at approximately 10 millimeters from the center of the sensing face with an operating distance of 3 millimeters. The sensor returns to its original state only after the target clears the release zone or release point. The difference (or distance) between the operating point and the release point is known as the *differential* (or hysteresis).

If extreme vibration is present, the user should ensure that the target clears the boundary of the release zone sufficiently. Stopping the target between the operating zone and the release zone could allow vibration to move the target enough to cause the sensor's output and load to *chatter*.

If a target other than a standard size target is used, the shape of the field map will be similar, but the overall dimensions will be reduced.

Helpful Application Hints

1. Avoid mounting locations where sensors are subject to direct contact or collision with the object being sensed.

2. If head-on (axial) operation is desired, an overtravel of only 25 percent should be allowed due to coasting of the target before it comes to a complete stop. For example, if the usable sensing range of a device is 4 millimeter, a target approaching the sensing face head on should stop approximately 3 millimeters from the sensing face to avoid collision.

3. If lateral (slide-by) operation is desired, the *tip* of the recommended sensing field should not be used, as variations in sensing distance are most significant in that area. Therefore, it is recommended that the target pass not more than 75 percent of the sensing distance from the sensor's face and not close enough to damage the sensor.

4. If a sensor is installed in an application where its cable must encounter continuous bending, make sure that there is enough slack in the cable to prevent broken conductors.

5. Electric cables (carrying DC) near the sensing face can affect operation if the field intensity is extremely high. If this exists, the conductors should be relocated away from the sensor's face. Also, when a sensor is located in a welding application, a weld, field-immune device should be considered.

CHAPTER

Control Transformers

Most industrial motors operate on voltages that range from 240 to 480 volts. Magnetic control systems, however, generally operate on 120 volts. A *control transformer* is used to step the 240 or 480 volts down to 120 volts to operate the control system. There is really nothing special about a control transformer except that most of them are made with two primary windings and one secondary winding. Each primary winding is rated at 240 volts, and the secondary winding is rated at 120 volts. This means there is a turns ratio of 2:1 (2 to 1) between each primary winding and the secondary winding. For example, assume that each primary winding contains 200 turns of wire and that the secondary winding contains 100 turns. There are two turns of wire in each primary winding for every one turn of wire in the secondary.

One of the primary windings of the control transformer is labeled H_1 and H_2. The other primary winding is labeled H_3 and H_4. The secondary winding is labeled X_1 and X_2. If the transformer is to be used to step 240 volts down to 120 volts, the two primary windings are connected in parallel with each other.

If the transformer is to be used to step 480 volts down to 120 volts, the primary windings are connected in series. With the windings connected in series, the primary winding now has a total of 400 turns of wire, which makes a turns ratio of 4:1. When 480 volts are connected to the primary winding, the secondary winding has an output of 120 volts.

Control transformers generally have screw terminals connected to the primary and secondary leads. If the transformer is to be connected for 240-volt operation, the two primary windings must be connected in parallel with each other. This connection can be made on the transformer by using one metal link to connect leads H_1 and H_3 and another metal link to connect H_2 and H_4.

If the transformer is to be used for 480-volt operation, the primary windings must be connected in series. This connection can be made on the control transformer by using a metal link to connect H_2 to H_3. Figure 8–1 is a control transformer used to step down 480 volts to 24 volts.

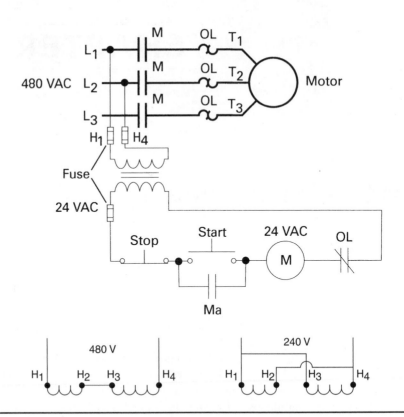

Figure 8–1   Control transformer used to step down voltage for control application. Courtesy Siemens Energy and Automation, Inc.

CHAPTER 9

Solenoid Valves

Valves are mechanical devices designed to control the flow of fluids such as oil, water, air, and other gases. Many valves are manually operated, but electrically operated valves are most often used in industry because they can be placed close to the devices they operate, thus minimizing the amount of piping required. Remote control is accomplished by running a single pair of control wires between the valve and a control device such as a manually operated switch or an automatic device.

A *solenoid valve* is a combination of two basic units: an assembly of the solenoid (the electromagnet) and plunger (the core), and a valve containing an opening in which a disk or plug is positioned to regulate the flow. The valve is opened or closed by the movement of the magnetic plunger. When the coil is energized, the plunger (core) is drawn into the solenoid (electromagnet). The valve operates when current is applied to the solenoid. The valve returns automatically to its original position when the current ceases.

Most control pilot devices operate a single-pole switch, contact, or solenoid coil. The wiring diagrams of these devices are not difficult to understand, and the actual devices can be connected easily into systems. It is recommended that the electrician know the purpose of and understand the action of the total industrial system for which various electric-control elements are to be used. In this way, the electrician will find it easier to design or assist in designing the electric-control system. It will also be easier for the electrician to install and maintain the control system.

Two-Way Solenoid Valves

Two-way (in-and-out) solenoid valves are magnetically operated valves that are used to control the flow of Freon, methyl chloride, sulphur dioxide, and other liquids in refrigeration and air-conditioning systems. These valves can also be used to control the flow of water, oil, and air.

Standard applications of solenoid valves generally require that the valve be mounted directly in line in the piping with the inlet and outlet connections directly opposite each other. Simplified valve mounting is possible with the use of a bottom outlet, which eliminates elbows and bends. In the bottom-outlet arrangement, the normal, side outlet is closed with a standard pipe plug.

The valve body is usually a special brass forging, which is carefully checked and tested to ensure that there will be no seepage due to parasites. The armature, or plunger, is made from a high-grade stainless steel. The effects of residual magnetism are eliminated by the use of a kickoff pin and spring that prevent the armature from sticking. A shading coil ensures that the armature will make a complete seal with the flat surface above it to eliminate noise and vibration.

It is possible to obtain DC coils with a special winding that will prevent the damage that normally results from an instantaneous voltage surge when the circuit is broken. Surge capacitors are not required with this type of coil.

To ensure that the valve will always seat properly, it is recommended that strainers be used to prevent grit or dirt from lodging in the orifice or valve seat. Dirt in these locations can cause leakage. The inlet and outlet connections of the valve must not be reversed. The tightness of the valve depends to a degree on pressure acting downward on the sealing disk. This pressure is possible only when the inlet is connected to the proper point as indicated on the valve.

Four-Way Solenoid Valves

Electrically operated, four-port, four-way air valves are used to control a double-acting cylinder (Figure 9–1). When the coil is deenergized, one side of the piston is at atmospheric pressure and the other side is acted upon by the line pressure. When the valve-magnet coil is energized, the valve exhausts the high-pressure side of the piston to atmospheric pressure. As a result, the piston and its associated load reciprocate in response to the valve movement. Four-way valves are used extensively in industry to control the operation of the pneumatic cylinders used on spot welders, press clutches, machine and assembly jig clamps, tools, and lifts.

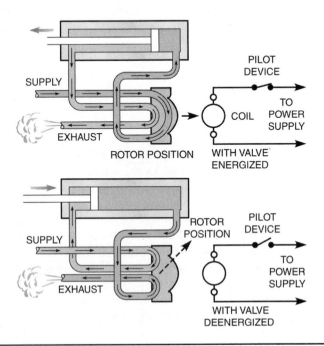

Figure 9–1   Control of double-acting cylinder by a four-way, electrically operated valve shown with elementary diagrams.

CHAPTER 10

Control Circuits

Developing Schematics and Wiring Diagrams[1]

Schematic and wiring diagrams are the written language of control circuits. If a maintenance electrician is going to install control equipment or troubleshoot existing control circuits, he must be able to interpret schematic and wiring diagrams.

Schematic diagrams, also known as line diagrams and ladder diagrams, show components in their electric sequence without regard to physical location. Schematics (see Figure 10–1A) are used more than any other type of diagram to connect or troubleshoot a control circuit.

Wiring diagrams show a picture of the control components with connecting wires. Wiring diagrams (see Figure 10–1B) are sometimes used to install new control circuits; they are seldom used for troubleshooting existing circuits.

Schematic Diagrams[2]

The method of expressing the language of control symbols is the *schematic diagram.* Schematic diagrams are made up of two circuits, the control circuit and the power circuit. Electric wires in a schematic diagram are represented by lines. Control-circuit wiring is represented by a lighter-weight line, and power-circuit wiring is represented by a heavier-weight line. A small dot or node at the intersection of two or more wires indicates an electric connection.

Schematic diagrams show the *functional* relationship of components and devices in an electric circuit, *not* the physical relationship. Figure 10–2 symbolically illustrates the functional relationship of a push button and a pilot light in a schematic diagram.

Schematic diagrams are *read* from top to bottom and from left to right.

1. Read a schematic as you would a book—from top to bottom and from left to right.

2. Contact symbols are shown in their deenergized or off position.

[1]Adapted with permission from Herman, *Industrial Motor Control 4,* Delmar Publishers.

[2]Adapted with permission from Siemens Energy and Automation, Inc., *Basics of Control Components,* Delmar Publishers.

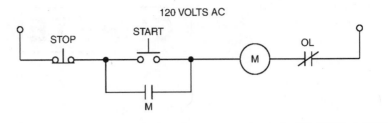

Figure 10–1A Stop/start schematic diagram.

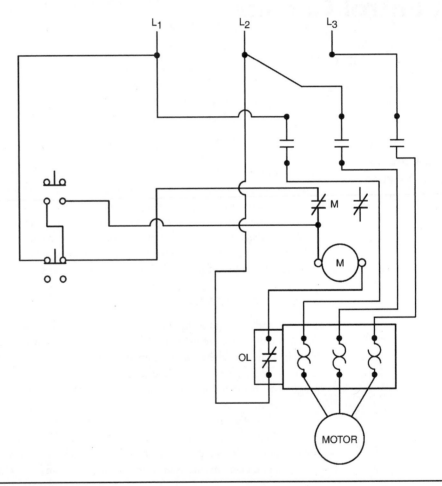

Figure 10–1B Wiring diagram of a start-stop, push-button station.

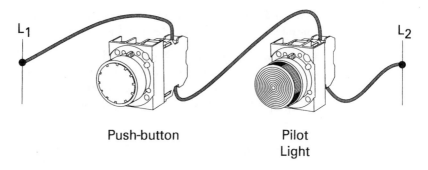

Figure 10–2 Functional relationship of push buttons. Courtesy Siemens Energy and Automation, Inc.

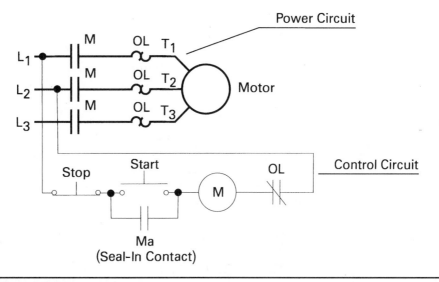

Figure 10–3 Single phase control wiring control and three phase power to motor. Courtesy Siemens Energy and Automation, Inc.

3. When a relay is energized, all the contacts controlled by that relay change position. If a contact is shown normally open on the schematic, it will close when the coil controlling it is energized.

Depressing the push button would allow current to flow from L_1 through the push button, illuminating the pilot light, and returning to L_2. Releasing the push button stops current flow, turning the pilot light off.

Power Circuit and Control Circuit   The *power circuit,* indicated by the heavier-weight line, is what actually distributes power from the source to the connected load (motor). The *control circuit,* indicated by the lighter-weight line, is used to *control* the distribution of power (see Figure 10–3).

Control Loads and Control Devices   Control circuits are made up of control loads and control devices. The *control load* is an electric device that uses electric power. Pilot lights, relays, and contacts are examples of control loads. *Control devices* are used to activate the control load. Push buttons and switches are examples of control devices.

Connecting Control Devices Control devices are connected between L_1 and the load. The control device can be connected in series or in parallel, depending on the desired results. In Figure 10–4, the push buttons are connected in parallel. Depressing either push button will allow current to flow from L_1, through the depressed push button, through the pilot light, to L_2.

Wiring Diagrams*

By superimposing the power-circuit diagram over the control-circuit diagram (Figure 10–5), a composite picture of all points of connection on the magnetic starter can be obtained. The

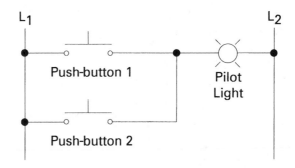

Figure 10–4 Devices can be connected in series or in parallel. Courtesy Siemens Energy and Automation, Inc.

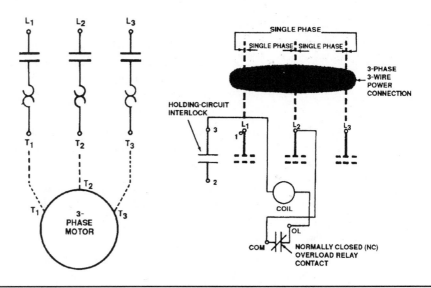

Figure 10–5 Power-circuit (left) and control-circuit (right) diagrams. Courtesy Square D Company.

*Adapted with permission from Square D Company.

combined illustration (Figure 10–6) is identified as a *wiring diagram.* It shows, as closely as possible, the actual location of all of the component parts of the starter. The dotted lines represent power-circuit connections made to the starter by the user.

Since wiring connections and terminal markings are shown, this type of diagram is helpful when wiring the starter or tracing wires when troubleshooting. Note that bold lines denote the power circuit and thin lines are used to show the control circuit. Conventionally, in AC magnetic equipment, black wires are used in power circuits and red wires are used for control circuits.

A wiring diagram, however, is limited in its ability to convey a clear picture of the sequence of operation of a controller. Where an illustration of the circuit in its simplest form is desired, the elementary diagram is used.

Elementary Diagrams

The *elementary diagram* (Figure 10–7) gives a simple picture of the circuit. The devices and components are not shown in their actual positions. All of the control-circuit components are shown as directly as possible between a pair of vertical lines representing the control power supply. The arrangement of the components is designed to show the sequence of operation of the devices and helps in understanding how the circuit operates. The effect of operating various interlocks and control devices can be readily seen, which helps in troubleshooting, particularly with the more complex controllers. This form of electrical diagram is sometimes referred to as a *schematic* or *one-line* diagram.

Two-Wire Control

In Figure 10–7, two wires connect the control device (which could be a thermostat, float switch, limit switch, or other maintained-contact device) to the magnetic starter. When the contacts of the control device close, they complete the coil circuit of the starter, causing it

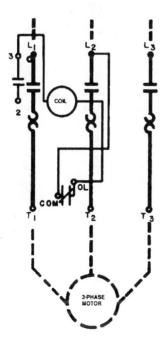

Figure 10–6 Wiring diagram. Courtesy Square D Company.

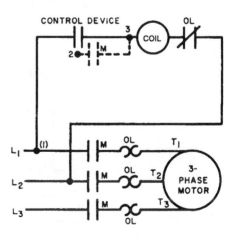

Figure 10–7 Elementary diagram of starter with two-wire control. Courtesy Square D Company.

to pick up and connect the motor to the power source. When the control-device contacts open, the starter is deenergized, stopping the motor.

Two-wire control provides low-voltage release, but not low-voltage protection. Wired as illustrated, the starter will function automatically in response to the control device without the attention of an operator.

The dotted portion shown in the elementary diagram in Figure 10–8 represents the holding-circuit interlock furnished on the starter, but not used in two-wire control. For greater simplicity, this portion is omitted from the conventional, two-wire elementary diagram.

Three-Wire Control

A three-wire control circuit uses momentary-contact *START-STOP* buttons and a holding-circuit interlock wired in parallel with the *START* button to maintain the circuit (Figure 10–9). Pressing the normally open *START* button completes the circuit to the

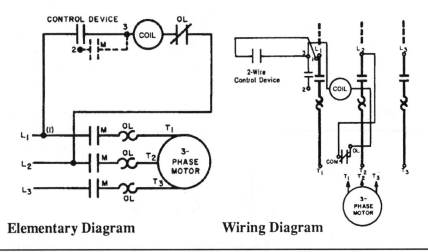

Elementary Diagram   **Wiring Diagram**

Figure 10–8   Diagram of a starter with two-wire control. Courtesy Square D Company.

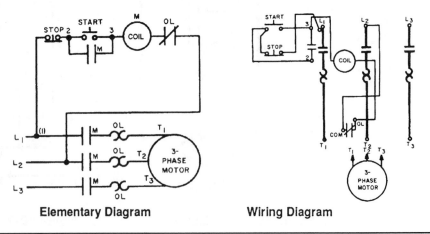

Elementary Diagram   **Wiring Diagram**

Figure 10–9   Diagram of a starter with three-wire control. Courtesy Square D Company.

coil. The power-circuit contacts in L_1, L_2, and L_3 close, completing the circuit to the motor, and the holding-circuit contact (mechanically linked with the power contacts) also closes. Once the starter has picked up, the *START* button can be released, as the now closed contact provides an alternate current path around the reopened start contact.

Pressing the normally closed *STOP* button will open the circuit to the coil, causing the starter to drop out. An overload condition, which causes the overload contact to open, a power failure, or a drop in voltage to less than the seal-in value, would also deenergize the starter. When the starter drops out, the interlock contact reopens, and both current paths to the coil—through the *START* button and the interlock—are now open. Since three wires from the push-button station are connected into the starter (at points *1, 2,* and *3*) this wiring scheme is commonly referred to as three-wire control.

Understanding Control Components and Symbols

Electrical Symbols

A language of controls has been developed to transfer an understanding of ideas and information. The language consists of a commonly used set of symbols that represent control components: contact symbols, switch symbols, push-button symbols, coil symbols, overload relay symbols, pilot-light symbols, transformers, inductors, and resistors.

Contact Symbols

Contact symbols are used to indicate an open or closed path of current flow. Contacts are shown as *NO* (normally open) or *NC* (normally closed). Contacts shown by these symbols require another device to actuate them.

The standard method of showing a contact is by indicating the circuit condition it produces when the actuating device is in the deenergized or nonoperated state. For example, in Figure 10–10, the relay is used as the actuating device. The contacts are shown as normally open, meaning the contacts are open when the relay is deenergized. A complete path of current does not exist, and the light is off.

Normally Open Contact   In a control diagram or schematic, symbols are usually not shown in the energized or operated state. For the purposes of explanation in this chapter, a contact or device shown in a state opposite of its normal state will be highlighted when in a schematic or wiring diagram. For example, in Figure 10–10, the circuit is first shown in the deenergized state. The contacts are shown in their *normally open* (*NO*) state. When a relay is energized, the contacts close, completing the path of current. The contacts have been highlighted to indicate that they are now closed in an energized state.

Normally Closed Contact   In Figure 10–10, the contacts are shown as *normally closed* (*NC*), indicating that the contacts are closed when the relay is deenergized. A complete path of current exists. When the relay is energized, the contacts open. The contacts have been highlighted to show that they are now open in an energized state.

Switch Symbols

Switch symbols are also used to indicate an open or closed path of current flow. Variations of this symbol are used to represent limit switches, foot switches, pressure switches, level switches, temperature-actuated switches, flow switches, and selector switches. Switches, like contacts, require another device or action to change their state. In the case of a manual switch, someone must manually change the position of the switch. See Appendix E for symbol descriptions.

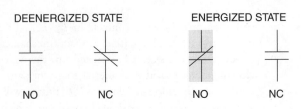

Figure 10–10 *NO,* normally opened, and *NC,* normally closed. Courtesy Siemens.

Push-Button Symbols

There are two basic types of *push-buttons:* momentary and maintained. A normally open momentary push button closes as long as the button is held down. A normally closed momentary push button opens as long as the button is held down. A maintained push button latches in place when the button is pressed. See Appendix E.

Coil Symbols

Coils are used in electromagnetic starters, contacts, and relays. The purpose of contacts and relays is to open and close associated contacts. A letter is used to designate the coil; for example, *M* frequently indicates a motor starter, and *CR* indicates a control relay. The associated contacts have the same identifying letter. Contacts and relays use an electromagnetic action, which is described later, to open and close these contacts. The associated contacts can be either normally open or normally closed. See Appendix E.

Overload Relay Symbols

Overload relays are used to protect motors from overheating due to an overload on the driven machinery, low line voltage, or an open phase in a three-phase system. When excessive current is drawn for a predetermined amount of time, the relay opens and the motor is disconnected from its source of power. See Appendix E.

Pilot-Light Symbols

A *pilot light* is a small electric light used to indicate a specific condition of a circuit. For example, a red light might be used to indicate that a motor is running. The letter in the center of the pilot light symbol indicates the color of the light. See Appendix E.

CHAPTER 11

Direct Current Motor Control

Small DC motors can be connected directly across the line for starting because a small amount of friction and inertia is overcome quickly in gaining full speed and developing a counter-electromotive force. Fractional-horsepower manual starters or magnetic contacts and starters are used for across-the-line starting of small DC motors.

Magnetic across-the-line control of small DC motors is similar to AC control or to two- or three-wire control. Some DC across-the-line starter coils have dual windings because of the added load of multiple-break contacts and the fact that the DC circuit lacks the inductive reactance that is present with AC electromagnets. Both windings are used to lift and close the contacts, but only one winding remains in the holding position. The starting (or lifting) winding of the coil is designed for momentary duty only.

Definite-Time Starting Control

When large DC motors are to be started using a *definite-time starting control,* current inrush to the armature must be limited. One method of limiting this current is to connect resistors in series with the armature. When the armature begins to turn, counter-electromotive force is developed in the armature. As counter-electromotive force increases, resistance can be shunted out of the armature circuit, permitting the armature to turn at a higher speed. When armature speed increases, counter-electromotive force also increases. Resistance can be shunted out of the circuit in steps until the armature is connected directly to the powerline.

Limiting the starting current of the armature is not the only factor that should be considered in a DC control circuit. Most DC motor-control circuits use a *field-current relay* (*FCR*) connected in series with the shunt field of the motor. The field-current relay ensures that current is flowing through the shunt field before voltage can be connected to the armature.

If the motor is running and the shunt field opens, the motor will become a series motor and begin to increase rapidly in speed. If this happens, both the motor and the equipment it is operating can be destroyed. For this reason, the shunt-field relay must disconnect the armature from the line if shunt-field current stops flowing.

Time-Delay Starting

The circuit shown in Figure 11–1 is a DC motor control with two steps of resistance connected in series with the armature. When the motor is started, both resistors limit current flow to the armature. *Time-delay* relays are used to shunt the starting resistors out of the circuit in time intervals of 5 seconds each until the armature is connected directly to the line.

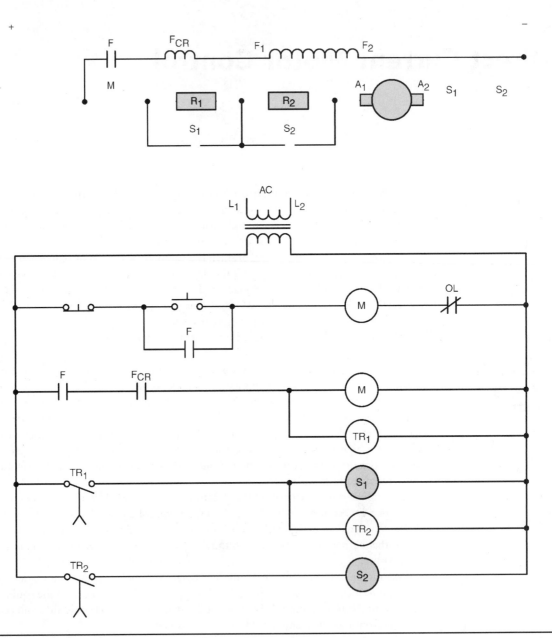

Figure 11–1   Time-delay starter for a DC motor.

The circuit operates as follows:

1. When the start button is pushed, current is supplied to relay coil F, and all F contacts change position. One F contact is connected in parallel with the start button and acts as a holding contact. Another F contact connects the field-current relay and the shunt field to the line.

2. When shunt-field current begins to flow, contact F_{CR} closes. When contact F_{CR} closes, a circuit is completed to motor-starter coil M and coil TR_1.
    a. When starter M energizes, contact M closes and connects the armature circuit to the DC line.
    b. Five seconds after coil TR_1 energizes, contact TR_1 closes. This permits current to flow to relay coils S_1 and TR_2.
    c. When contact S_1 closes, resistor R_1 is shunted out of the circuit.
    d. Five seconds after relay coil $TR2$ energizes, contact TR_2 closes. When contact TR_2 closes, current can flow to coil S_2.
    e. When contact S_2 closes, resistor R_2 is shunted out of the circuit and the armature is connected directly to the DC powerline.

3. When the stop button is pushed, relay F deenergizes and opens all F contacts. This breaks the circuit to starter coil M, which causes contact M to open and disconnect the armature from the line.

4. When coil TR_1 deenergizes, contact TR_1 opens immediately and deenergizes coils S_1 and TR_2. When coil TR_2 deenergizes, contact TR_2 opens immediately and deenergizes coil S_2. All contacts in the circuit are back in their original positions, and the circuit is ready to be started again.

Using Solid-State for DC Motor Controls

Direct current motors are used throughout much of industry because of their ability to produce high torque at low speed, and because of their variable-speed characteristics. Direct current motors are generally operated at or below normal speed. *Normal speed* for a DC motor is obtained by operating the motor with full rated voltage applied to the field and armature. The motor can be operated at below normal speed by applying rated voltage to the field and reduced voltage to the armature.

In Figure 11–1, resistance was connected in series with the armature to limit current and, therefore, speed. Although this method does work and was used in industry for many years, it is seldom used today. When resistance is used for speed control, much of the power applied to the circuit is wasted in heating the resistors, and the speed control of the motor is not smooth because resistance is taken out of the circuit in steps.

Speed control of a DC motor is much smoother if two separate power supplies, which convert the AC voltage to DC voltage, are used to control the motor instead of resistors connected in series with the armature. These two power supplies are the shunt-field power supply and the armature power supply.

The Shunt-Field Power Supply

Most solid-state DC motor controllers provide a separate DC power supply that is used to furnish excitation current to the shunt field. The *shunt field* of most industrial motors requires a current of only a few amperes to excite the field magnets; therefore, a small power supply can be used to fulfill this need. The *shunt-field power supply* is generally designed to remain turned on even when the main (armature) power supply is turned off. If power is connected to the shunt field even when the motor is not operating, the shunt field will act as a small resistance heater for the motor. This heat helps prevent moisture from forming in the motor due to condensation.

The Armature Power Supply

The *armature power supply* is used to provide variable DC voltage to the armature of the motor. This power supply is the heart of the solid-state motor controller. Depending on the size and power rating of the controller, armature power supplies can be designed to produce from a few amperes to hundreds of amperes. Most of the solid-state motor controllers intended to provide the DC power needed to operate large DC motors convert three-phase AC voltage directly into DC voltage with a three-phase bridge rectifier.

The diodes of the rectifier, however, are replaced with SCRs to provide control of the output voltage. A large diode is often connected across the output of the bridge. This diode is known as a *freewheeling* or *kickback diode* and is used to kill inductive spike voltages produced in the armature. If armature power is suddenly interrupted, the collapsing magnetic field induces a high voltage into the armature windings. The diode is reverse biased when the power supply is operating under normal conditions, but an induced voltage is opposite in polarity to the applied voltage. This means the kickback diode will be forward biased to any voltage induced into the armature. Since a silicon diode has a voltage drop of .6 to .7 volts in the forward direction, a high voltage spike cannot be produced in the armature.

Voltage Control

Output *voltage control* is achieved by phase shifting the SCRs. The phase-shift control unit determines the output voltage of the rectifier (Figure 11–2). Since the phase-shift unit is the real controller of the circuit, other sections of the circuit provide information to the phase-shift control unit.

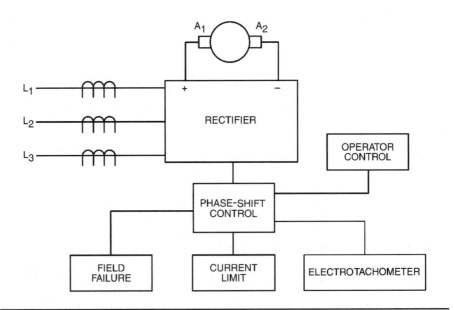

Figure 11–2 Electrotachometer measures motor speed.

Field-Failure Control

As stated previously, if current flow through the shunt field is interrupted, a compound-wound DC motor will become a series motor and race to high speeds. Some method must be provided to disconnect the armature from the circuit in case current flow through the shunt field stops. Several methods can be used to sense current flow through the shunt field. For example, a contact of the current relay is connected in series with the coil of a motor starter used to connect the armature to the powerline. If current flow is stopped, the contact of the current relay opens, causing the circuit of the motor-starter coil to open.

Another method used to sense current flow is to connect a low value of resistance in series with the shunt field. The voltage drop across the sense resistor is proportional to the current flowing through the resistor ($E = I \times R$). Since the sense resistor is connected in series with the shunt field, the current flow through the sense resistor must be the same as the current flow through the shunt field. A circuit can be designed to measure the voltage drop across the sense resistor. If this voltage falls below a certain level, a signal is sent to the phase-shift control unit and the SCRs are turned off (Figure 11–2).

Current-Limit Control

The armature of a large DC motor has a very low resistance, typically less than one ohm. If the controller is turned on with full voltage applied to the armature, or if the motor stalls while full voltage is applied to the armature, a very large current will flow. This current can damage the armature of the motor or the electronic components of the controller. For this reason, most solid-state, DC motor controls use some method to limit the current to a safe value.

One method of sensing the current is to insert a low value of resistance in series with the armature circuit. The amount of voltage dropped across the sense resistor is proportional to the current flow through the resistor. When the voltage drop reaches a certain level, a signal is sent to the phase-shift control telling it not to permit any more voltage to be applied to the armature (see Figure 11–2).

Speed Control   A common method of *controlling speed* is with an *electrotachometer*. The output voltage of the generator is proportional to its speed. The output voltage of the generator is connected to the phase-shift control unit (Figure 11–2). If load is added to the motor, the motor speed will decrease. When the motor speed decreases, the output voltage of the electrotachometer drops. The phase-shift unit detects the voltage drop of the tachometer and increases the armature voltage until the tachometer voltage returns to the proper value.

If the load is removed, the motor speed will increase. An increase in motor speed causes an increase in the output voltage of the tachometer. The phase-shift unit detects the increase of tachometer voltage and causes a decrease in the voltage applied to the armature. Electronic components respond so fast that there is almost no noticeable change in motor speed when load is added or removed.

Alternating Current Motor Control

Typical Starting Methods

The most common methods of starting polyphase squirrel cage motors include:

- *Full voltage starting:* a hand-operated or automatic starting switch throws the motor directly across the line.
- *Primary resistance starting:* a resistance unit connected in series with the stator reduces the starting current.
- *Autotransformer or compensator starting:* manual or automatic switching between the taps of the autotransformer gives reduced voltage starting.
- *Impedance starting:* reactors are used in series with the motor.
- *Star-delta starting:* the stator of the motor is star connected for starting and delta connected for running.
- *Part winding starting:* the stator windings of the motor are made up of two or more circuits; the individual circuits are connected to the line in series for starting and in parallel for normal operation.

Of these methods, the two most fundamental methods of starting squirrel cage motors are full-voltage starting and reduced-voltage starting. Once in, full-voltage starting can be used where the driven load can stand the shock of starting and objectionable line disturbances are not created. Reduced-voltage starting may be required if the starting torque must be applied gradually or if the starting current produces objectionable line disturbances.

Primary Resistor-Type Starters

A simple and common method of starting a motor at reduced voltage is used in primary resistor-type starters. In this method, a resistor is connected in series in the lines to the motor. Thus, there is a voltage drop across the resistors and the voltage is reduced at the motor terminals. Reduced-motor starting speed and current are the result. As the motor accelerates, the current through the resistor decreases, reducing the voltage drop ($E = IR$) and increasing the voltage across the motor terminals. A smooth acceleration is obtained with gradually increasing torque and voltage.

The resistance is disconnected when the motor reaches a certain speed. The motor is then connected to run on full line voltage. The introduction and removal of resistance in the motor starting circuit may be accomplished manually or automatically.

Primary resistor starters are used to start squirrel cage motors in situations where limited torque is required to prevent damage to driven machinery. These starters are also used with limited current inrush to prevent excessive power line disturbances.

It is desirable to limit the starting current in the following cases:

1. When the power system does not have the capacity for full-voltage starting.
2. When full-voltage starting may cause serious line disturbances, such as in lighting circuits, electronic circuits, the simultaneous starting of many motors, or if the motor is distant from the incoming power supply.

In these situations, reduced-voltage starters may be recommended for motors with ratings as small as five horsepower.

Reduced voltage starting must be used for driving machinery that must not be subjected to a sudden high starting torque and the shock of sudden acceleration. Among typical applications are those where belt drives may slip or where large gears, fan blades, or couplings may be damaged by sudden starts.

Automatic primary resistor starters may use one or more steps of acceleration, depending upon the size of the motor being controlled. These starters provide smooth acceleration without the line current surges normally experienced when switching autotransformer types of reduced-voltage starters.

Primary resistor starters provide closed transition starting. This means the motor is never disconnected from the line from the moment it is first connected until the motor is operating at full line voltage. This feature may be important in wiring systems sensitive to voltage changes. Primary resistor starters do consume energy, with the energy being dissipated as heat. However, the motor starts at a much higher power factor than with other starting methods.

Special starters are required for very high inertia loads with long acceleration periods or where power companies require that current surges be limited to specific increments at stated intervals.

Primary Resistor-Type Reduced Voltage Starter

Figure 12–1 illustrates an automatic, primary resistor-type, reduced-voltage starter. The starter is shown with two-point acceleration connected to a three-phase, squirrel cage induction motor.

When the start button is pressed, a complete circuit is established beginning at L_1 and continuing through the stop button, start button, and coil M, until the overload relay contacts to L_2. When coil (M) is energized, the main power contacts (M) and the control circuit maintaining contact (M) are closed. The motor is energized through the overload heaters and the starting resistors. Because the resistors are connected in series with the motor terminals, a voltage drop occurs in the resistors and the motor starts on reduced voltage.

As the motor accelerates, the voltage drop across the resistors decreases gradually because of reduction in the starting current. At the same time, the motor terminal voltage increases.

After a predetermined acceleration time, delay contact M closes the circuit to contactor coil S. Coil S, in turn, closes contacts S, the resistance shunted out, and the motor is connected across the full line voltage.

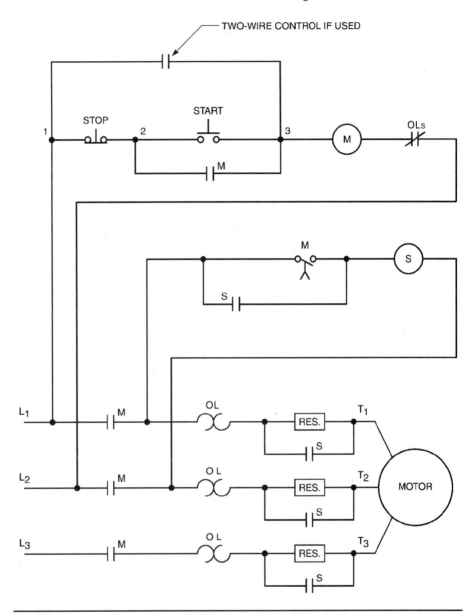

Figure 12-1 Line diagram of a primary resistor starter with two-step acceleration.

Note that the stop button controls coil *M* directly. When the main power contacts *M* open, coil *S* drops out. After coil *M* is energized, a pneumatic timing unit attached to starter unit *M* reads the closing of time contact *M*. This scheme uses starter *M* for a dual purpose and eliminates the coil, a timing relay. Additional control contact (*S*) connected parallel to the on-delay contact (*M*) provides additional electrical security to insure coil(s) normally open is maintained.

For maximum operating efficiency, push buttons or other pilot devices are usually mounted on the driven machinery within easy reach of the operator. The starter is located near the motor to keep the heavy power circuit wiring as short as possible. Only two or three small connecting wires are necessary between the starter and pilot device.

A motor can be operated from any of several remote locations if a number of push buttons or pilot switches are used with one magnetic starter, such as on a conveyor system.

Alternating current primary resistor starters are available for use on single-phase and three-phase reversing operations. They are also available with multiple points of acceleration.

A primary resistance-type starter has the following features:

- Simple construction
- Low initial cost
- Low maintenance
- Smooth acceleration in operation
- Continuous connection of the motor to the line during the starting period
- High power factor

These starters should not be used for starting very heavy loads because of their low starting torque. These starters have a low starting economy because the starting resistors dissipate electrical energy.

Autotransformer Starters

Autotransformer reduced-voltage starters are similar to primary resistor starters in that they are used primarily with AC squirrel cage motors to limit the inrush current or to lessen the starting strain on driven machinery. This type of starter uses autotransformers between the motor and the supply lines to reduce the motor-starting voltage. Taps are provided on the autotransformer to permit the user to start the motor at approximately 50 percent, 65 percent, or 80 percent of line voltage.

Most motors are successfully started at 65 percent of line voltage. In situations where this value of voltage does not provide sufficient starting torque, the 80 percent tap is available. If the 50 percent starting voltage creates excessive line drop to the motor, the 65 percent tap is available. This way of changing the starting voltage is not usually available with other types of starters. The starting transformers are inductive loads; therefore, they momentarily affect the power factor. They are suitable for long starting periods, however.

To reduce the voltage across the motor terminals during the accelerating period, an autotransformer-type starter generally has two autotransformers connected in open delta. During the reduced voltage starting period, the motor is connected to the taps on the autotransformer. With the lower starting voltage, the motor draws less current and develops less torque than if it were connected to the line voltage.

An adjustable time-delay relay controls the transfer from the reduced voltage condition to full voltage. A current-sensitive relay may be used to control the transfer to obtain current-limiting acceleration.

To understand the operation of the autotransformer starter more clearly, refer to the line diagram in Figure 12–2. When the start button is closed momentarily, the timing relay (TR) is energized. The relay maintains the circuit across the start button with the normally open instantaneous contact (TR) which now closes. Starting coil (S) is energized from terminal four through the normally closed "time delay in opening" contact (TR), through the normally closed interlock (R), through coil S, and through the overload contact to L_2, com-

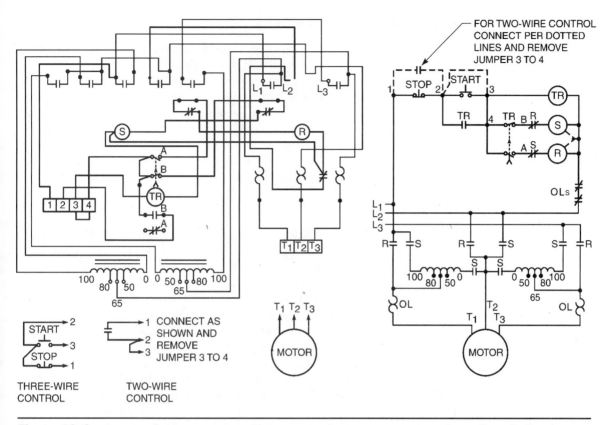

Figure 12–2 Autotransformer starters provide greater starting torque per ampere drawn from the line than any other type of reduced voltage starter. A typical wiring diagram (left) and a line diagram (right).

pleting the circuit. The running starter cannot be closed at this point because the normally closed interlock *S* is open and the mechanical interlock is operating.

After a preset timing period (*TR*), the normally closed contacts open and the normally open (*TR*) contacts close. When coil *S* is deenergized, normally closed interlock *S* closes and energizes the running starter *R*.

The contact switching arrangement for a typical power circuit is shown in Figure 12–2. When two transformers are used, there will be an imbalance in the motor voltage during starting. This imbalance will produce a torque variation of approximately 10 percent. In the running position, the motor is connected directly across the line and the autotransformers are disconnected from the line. As a result, only three contacts are shown.

Full line voltage is applied to the outside terminals of the autotransformer on starting. Reduced voltage for starting the motor is obtained from the autotransformer taps. The current taken by a motor varies directly with the applied voltage.

Starting compensators (autotransformer starters) using a five-pole starting contactor are classified as open transition starters. The motor is disconnected momentarily from the line during the transfer from the start to the run conditions.

Automatic Starters for Star-Delta Motors

A commonly used means of reducing inrush currents without the need of external devices is star-delta motor starting (sometimes called wye-delta starting).

Star-delta motors are similar in construction to standard squirrel cage motors. However, in star-delta motors, both ends of each of the three windings are brought out to the terminals. If the starter used has the required number of properly wired contacts, the motor can be started in star and run in delta.

The motor must be wound in such a manner that it will run with its stator windings connected in delta. The leads of all of the windings must be brought out to the motor terminals for their proper connection in the field.

Applications

The primary applications of star-delta motors are for driving centrifugal chillers of large, central air-conditioning units for loads such as fans, blowers, pumps, or centrifuges, and for situations where a reduced starting torque is necessary. Star-delta motors also may be used where a reduced starting current is required. Since all of the stator winding is used and there are no limiting devices such as resistors or autotransformers, star-delta motors are widely used on loads having high inertia and a long acceleration period.

The speed of a star-delta or wye-delta squirrel cage induction motor depends on the frequency of the applied voltage and the number of stator poles. Since both these values are the same for the wye or delta connection, the motor will run at approximately the same speed regardless of how the windings are connected. The inrush line current is much less when the windings are connected in wye, however. Assume that a motor is to be connected directly to a 480-volt line during the starting period. Also assume that each winding exhibits an impedance of 0.4 during the starting period. If the windings are connected in delta when power is first applied to the motor, 480 volts will be connected directly across the phase windings. This will produce a phase current of 1200 amps. This can be illustrated by the following:

$$I\ PHASE = \frac{E\ PHASE}{Z\ PHASE}$$
$$I\ PHASE = \frac{480}{0.4}$$
$$I\ PHASE = 1200\ amps$$

Since the line current supplying a delta connection is 1.732 times greater than the phase current, the line current will be 2078.4 amps ($1200 \times 1.732 = 2078.4$).

If the stator windings are connected in a wye configuration during the starting period, the inrush line current will be only one third the value of the delta connection. Since the windings are now connected in a wye or star, the voltage applied across each phase winding will be less than the line voltage by a factor of 1.732 or 277 volts ($480 \div 1.732 = 277$). This will produce a phase current of 692.8 amps when power is first connected to the motor, as illustrated by the following:

$$I\ PHASE = \frac{E\ PHASE}{Z\ PHASE}$$
$$I\ PHASE = \frac{277}{0.4}$$
$$I\ PHASE = 692.8\ amps$$

In a wye-connected system, the line current and phase current are the same. Therefore, the line current has been reduced from 2078.2 amps to 692.8 amps during the initial starting period.

Overload Protection

Three overload relays are connected in the phase windings during both the starting and running period (Figure 12–3). This means that the overload heaters must be selected on

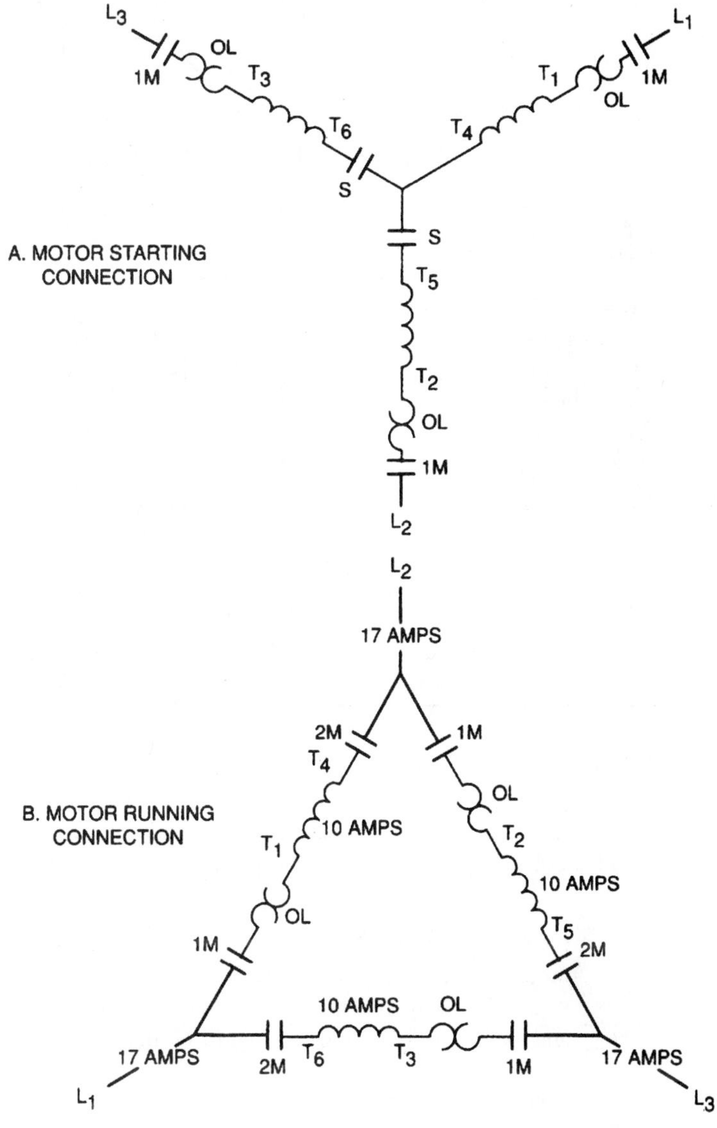

Figure 12–3 Elementary diagrams of motor power circuits. Controller connects motor in wye on start and in delta for run. Note that the overload relays are connected in the motor winding circuit, not in the line. Note also that the line current is higher than the phase winding current in the diagram for the delta connection (B). Winding current is the same as the line current in diagram A.

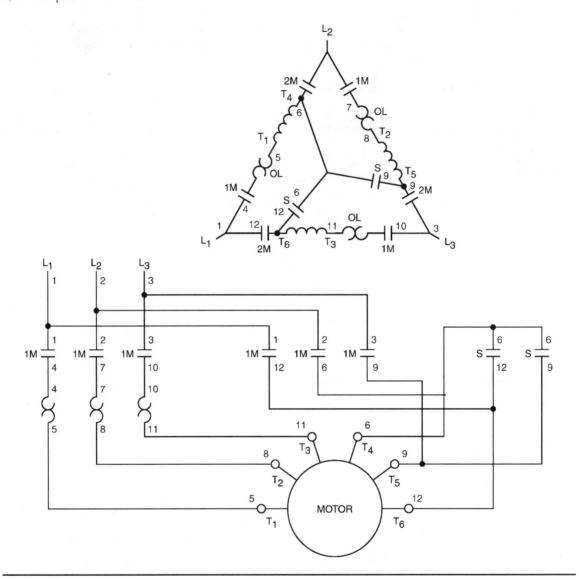

Figure 12–4 Load circuit connection for wye-delta starter.

the basis of the winding or phase current, not on the full-load line current indicated on the motor nameplate. To determine the proper current for the overload heaters, divide the line current by 1.732. A diagram of the entire connection for the motor windings, load heaters, and load contacts is shown in Figure 12–4.

Open Transition Starting

Probably the most common method for wye-delta starting is *open transition* starting. This thod receives its name from the fact that the motor windings are open during the transition period of changing the windings from a wye connection to a delta connection. A control circuit for performing this transition is shown in Figure 12–5. In this circuit, an "on"

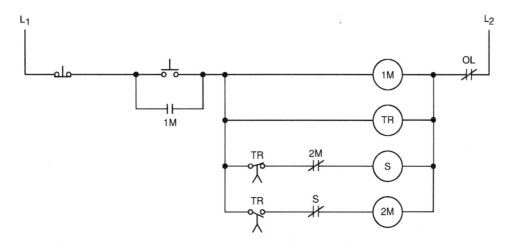

Figure 12–5   Basic control circuit for a wye-delta starter.

delay timer is used to change the motor windings from a wye connection to a delta connection. When the start button is pressed, coils *1M, TR,* and *S* energize immediately. Coil *1M* closes all *1M* contacts to supply power to the motor windings. Coil *S* closes both of the *S* contacts that connect the motor windings in a wye.

After some period of time, both *TR* contacts change position. The normally closed contact connected in a series with an *S* coil opens and deenergizes the *S* coil. The *S* load contacts open and disconnect the motor windings. When the normally open *TR* contact closes, coil *2M* energizes and reconnects the motor windings in a delta configuration. It is the transition period between *S* contacts opening and *2M* contacts closing that this starting method receives its name. The normally closed *S* and *2M* contacts act as interlocks to prevent the possibility of coils *S* and *2M* being energized at the same time.

Care should be exercised when connecting the stator windings to the load contacts. If the circuit is not connected properly, it will generally result in the motor reversing direction of rotation when the windings are changed from wye to delta. Figure 12–4 illustrates a schematic diagram of the stator winding connection and a wiring diagram of the motor connection to the load contacts. Notice that the wire numbers have been added to the stator schematic which corresponds to the components shown in the wiring diagram.

Closed Transition Starting

In Figure 12–6, resistors maintain continuity to the motor to avoid the difficulties associated with the open circuit form of transition between start and run. With closed transition starting, the transfer from the star to delta connections is made without disconnecting the motor from the line. When the transfer from star to delta is made in open transition starting, the starter momentarily disconnects the motor and then reconnects it in delta. While an open transition is satisfactory in many cases, some installations may require closed transition starting to prevent power line disturbances. Closed transition starting is achieved by adding a three-pole contactor and three resistors to the starter circuit. The connections are made as shown in the closed transition schematic diagram (Figure 12–6). The contactor is energized only during the transition from star to delta. It keeps the motor connected to the power source through the resistors during the transition period. There is a reduction in the

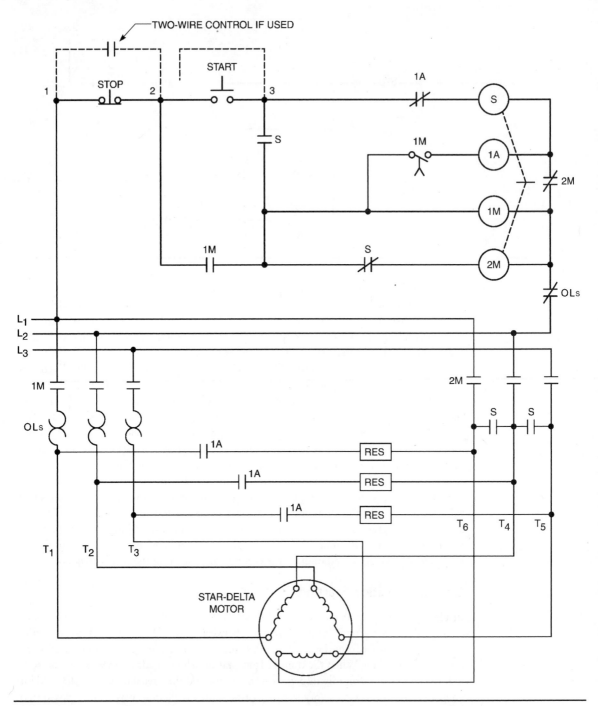

Figure 12–6 Elementary diagram of sizes 1, 2, 3, 4, and 5 star-delta starters with transition starting.

incremental current surge which results from the transition. The balance of the operating sequence of the closed transition starter is similar to that of the open transition star-delta motor starter.

A single method of reduced current starting may not achieve the desired results due to complicated motor starting requirements, stringent restrictions, and conflicting needs. It may be necessary to use a combination of starting methods before satisfactory performance is realized. For special installations, it may be necessary to design a starting system to fit the particular conditions.

Two-Speed, One-Winding (Consequent Pole) Motor Controller

Certain applications require the use of a squirrel cage motor having a winding arranged so that the number of poles can be changed by reversing some of the currents. If the number of poles is doubled, the speed of the motor is cut approximately in half.

The number of poles can be cut in half by changing the polarity of alternate pairs of poles. The polarity of half the poles can be changed by reversing the current in half the coils.

If a stator field is laid flat the established stator field must move the rotor twice as far in B as in A and in the same amount of time. As a result, the rotor must travel faster. The fewer the number of poles established in the stator, the greater the speed is in rpm of the rotor.

As shown in Figure 12–7A, a three-phase squirrel cage motor can be wound so that six leads are brought out. By making suitable connections with these leads, the windings can be connected in series delta or parallel wye (Figure 12–7B). If the winding is such that the series delta connection gives the high speed and the parallel wye connection gives the low speed, the horsepower rating is the same at both speeds. If the winding is such that the series delta connection gives the low speed and the parallel wye connection gives the high speed, the torque rating is the same at both speeds.

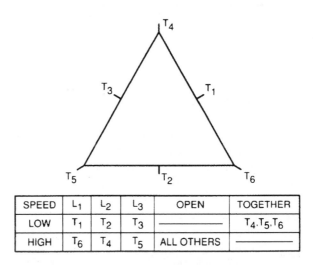

SPEED	L_1	L_2	L_3	OPEN	TOGETHER
LOW	T_1	T_2	T_3	———	$T_4.T_5.T_6$
HIGH	T_6	T_4	T_5	ALL OTHERS	———

Figure 12–7A   Connection table for a three-phase, two-speed, one-winding, constant horsepower motor.

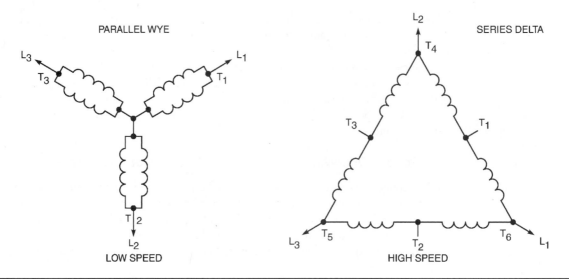

Figure 12–7B   Three-phase, two-speed, one-winding, constant horsepower motor connections made by motor controller.

Consequent pole motors have a single winding for two speeds. Extra taps can be brought from the winding to permit reconnection for a different number of stator poles. The speed range is limited to a 1:2 ratio, such as 600-1200 rpm or 900-1800 rpm.

Two-speed consequent pole motors have one reconnectable winding. However, three-speed consequent pole motors have two windings, one of which is reconnectable. Four-speed consequent pole motors have two reconnectable windings.

Referring back to the motor connection table in Figure 12–7A note that for low speed operation T_1 is connected to L_1; T_2 to L_2; T_3 to L_3; and T_4, T_5, and T_6 are connected together. For high-speed operation, T_6 is connected to L_1; T_4 to L_2; T_5 to L_3; and all other motor leads are open.

Figure 12–8 illustrates the circuit for a Size 1, selective multispeed starter connected for operation with a reconnectable, constant horsepower motor. The control station is a three-element, fast-slow-stop station connected for starting at either the fast or slow speed. The speed can be changed from fast to slow or slow to fast without pressing the stop button between changes. If equipment considerations make it desirable to stop the motor before changing speeds, this feature can be added to the control circuit by making connections D and A, represented in Figure 12–8 as dashed lines. Adding these jumper wires eliminates push-button interlocking in favor of stopping the motor between speed changes. This feature may be desirable in certain applications.

Connections for the addition of indicating lights or a two-wire pilot device instead of the control shown are also given.

Wound-Rotor Motors and Manual Speed Control

The AC three-phase *wound rotor,* or *slip ring,* induction motor was the first alternating-current motor that successfully provided speed control characteristics. This type of motor was an important factor in successfully adapting alternating current for industrial-power applications. Because of their flexibility in specialized applications, wound rotor motors and controls are widely used throughout the industry to drive conveyors for moving materials, hoists, grinders, mixers, pumps, variable speed fans, saws, and crushers. Advantages of this type of motor include maximum utilization of driven equipment,

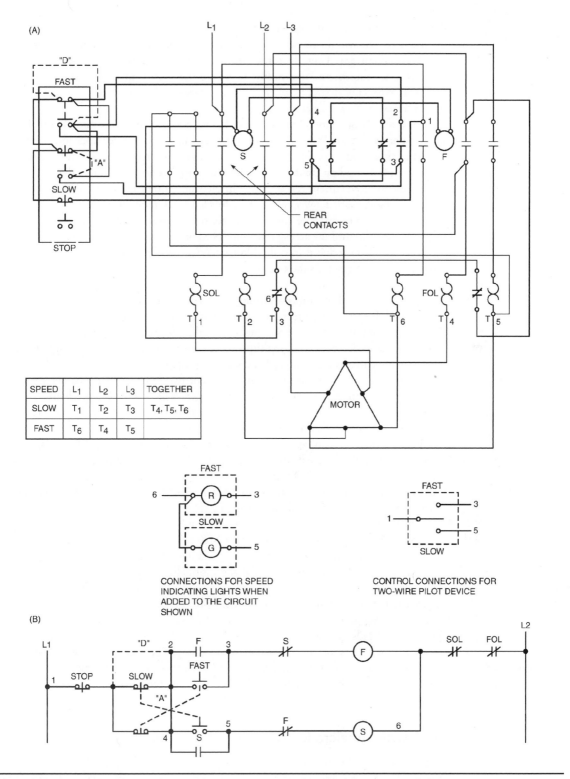

SPEED	L_1	L_2	L_3	TOGETHER
SLOW	T_1	T_2	T_3	T_4, T_5, T_6
FAST	T_6	T_4	T_5	

CONNECTIONS FOR SPEED
INDICATING LIGHTS WHEN
ADDED TO THE CIRCUIT
SHOWN

CONTROL CONNECTIONS FOR
TWO-WIRE PILOT DEVICE

Figure 12–8 Wiring diagram (A) and line diagram (B) of an AC, full voltage, two-speed magnetic starter for single-winding (reconnectable pole) motors.

better coordination with the overall power system, and reduced wear on mechanical equipment. The wound-rotor motor has the added features of high-starting torque and low-starting current. These features give the motor better operating characteristics for applications requiring a large motor or where the motor must start under load. This motor is especially desirable where its size is large with respect to the capacity of the transformers or power lines.

The phrase "wound rotor" actually describes the construction of the rotor. In other words, it is wound with wire. When the rotor is installed in a motor, three leads are brought out from the rotor winding to solid conducting slip rings. Carbon brushes ride on these rings and carry the rotor winding circuit out of motor to a controller. Unlike the squirrel cage motor, the induced current can be varied in the wound-rotor motor. As a result, the motor speed can also be varied. (Wound-rotor motors have stator windings identical to those used in squirrel cage motors.) The controller varies the resistance (and thus the current) in the rotor circuit to control the acceleration and speed of the rotor once it is operating.

Resistance is introduced into the rotor circuit when the motor is started or when it is operating at slow speed. As the external resistance is eliminated by the controller, the motor accelerates.

Generally, wound-rotor motors under load are not suitable for starting when the rings are shunted. If such a motor is started in this matter, the rotor resistance is so low that starting currents are too high to be acceptable. In addition, the starting torque is less than if suitable resistors are inserted in the slip-ring circuit. By inserting resistance in the ring circuit, starting currents are decreased and starting torque is increased.

A control for a wound-rotor motor consists of two separate elements: (1) a means of connecting the primary or stator winding to the power lines, and (2) a mechanism for controlling the resistance in the secondary or rotor circuit. For this reason, wound-rotor motor controllers are often called *secondary resistor starters.*

Basically, there are two types of manual controllers: starters and regulators. The resistors used in starters are designed for starting duty only. This means that the operating lever must be moved to the full ON position. The lever must not be left in any intermediate position. The resistors used in regulators are designed for continuous duty. As a result, the operating lever in regulators can be left in any speed position.

When wound-rotor motors are used as adjustable speed drives, they are operated on a continuous basis with resistance in the rotor circuit. In this case, the speed regulation of the motor is changed, and the motor operates at less than the full-load speed.

The use of three-wire control for starting (normally closed control contact) means that low-voltage protection is provided. The motor is disconnected from the line in the event of voltage failure. To restart the motor when the voltage returns to its normal value, the normal starting procedure must be followed. To reverse the motor, any two of the three-phase motor leads may be interchanged.

Drum controllers may be used for manual starting. The drum controller, however, is an independent component; it is separate from the resistors. A starting contact controls a line voltage, across-the-line starter. Generally, the electrician must make the connections on the job site.

Automatic Acceleration for Wound-Rotor Motors

Secondary resistor starters used for the automatic acceleration of wound-rotor motors consist of (1) an across-the-line starter for connecting the primary circuit to the line and

(2) one or more accelerating contactors to shunt out resistance in the secondary circuit as the rotor speed increases. The secondary resistance consists of banks of three uniform wye sections. Each section is to be connected to the slip rings of the motor. The wiring of the accelerating starters and the design of the resistor sections are meant for starting duty only. This type of controller cannot be used for speed regulation. The current inrush on starters with two steps of acceleration is limited by the secondary resistors to a value of approximately 250 percent at the point of the initial acceleration. Resistors on starters with three or more steps of acceleration limit the current inrush to 150 percent at the point of initial acceleration. Resistors for acceleration generally are designed to withstand one 10-second accelerating period in each 80 seconds of elapsed time, for a duration of one hour without damage.

The operation of accelerating contactors is controlled by a *timing device*. This device provides timed acceleration in a manner similar to the operation of primary resistor starters. Normally, the timing of the steps of acceleration is controlled by adjustable accelerating relays. When these timing relays are properly adjusted, all starting periods are the same regardless of variations in the starting load. This automatic timing feature eliminates the danger of an improper startup sequence by an inexperienced machine operator.

The primary circuit (stator) in Figure 12–9 is energized with the start button. The motor starts with a full value of resistance in the secondary circuit. Coil P actuates the normally open, delay-in-closing contact P. After a preset timing period, contact P closes, energizes contactor S_1, and maintains itself through the maintaining contact S_1. When contacts S_1 in the secondary resistor circuit close, the motor continues to accelerate. After the normally open, delay-in-closing contact S_1 times out and closes, contactor S_2 is energized and closes the resistor circuit contacts S_2. The motor then is accelerated to its maximum speed. The normally closed interlock S_2 opens the contactor (S_1) circuit. The closing of S_2 is assured by staggered, or overlapping, control contacts S_2.

Synchronous Motor Operation

One of the distinguishing features of the *synchronous motor* is that it runs without slip at a speed determined by the frequency of the connected power source and the number of poles it contains. This type of motor sets up a rotating field through stator coils energized by alternating current. (This action is similar to the principle of an induction motor.) An independent field is established by a rotor energized by direct current through slip rings mounted on the shaft. The rotor has the same number of coils as the stator. At running speed, these fields (north and south) lock into one another magnetically so that the speed of the rotor is in step with the rotating magnetic field of the stator. In other words, the rotor turns at the synchronous speed. Variations in the connected load do not cause a corresponding change in speed, as they would with the induction motor.

The rotor is excited by a source of direct current so that it produces alternate north and south poles. These poles are then attracted by the rotating magnetic field in the stator. The rotor must have the same number of poles as the stator winding. Every rotor pole, north and south, has an alternate stator pole, south and north, with which it can synchronize.

The rotor has DC field windings to which direct current is supplied through collector rings (slip rings). The current is provided from either an external source or a small DC generator connected to the end of the rotor shaft.

The magnetic fields of the rotor poles are locked into step with, and pulled around by, the revolving field of the stator. Assuming that the rotor and stator have the same number of poles, the rotor moves at the stator frequency (in hertz) actually produced by the generator supplying the motor.

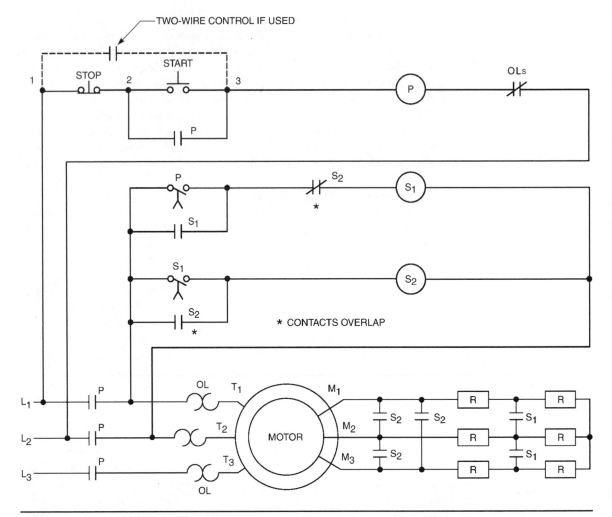

Figure 12–9 Typical elementary diagram of wound-rotor motor starter with three points of acceleration.

Synchronous motors are constructed similar to alternators. They differ only in those exact features of design that may make the motor better adapted to its particular purpose.

A synchronous motor cannot start without help because the DC rotor poles at rest are alternately attracted and repelled by the revolving stator field. Therefore, an induction squirrel cage, or starting winding, is embedded in the pole faces of the rotor. This is called an *amortisseur* (ah-more-tisir) winding.

This starting winding resembles a squirrel cage winding. The induction effect of the starting winding provides the starting, accelerating, and pull-up torques required. The winding is designed to be used only for starting and for damping oscillations during running. It cannot be used like the winding of the conventional squirrel cage motor. It has a relatively small cross-sectional area and will overheat if the motor is used as a squirrel cage induction motor.

The slip is equal to 100 percent at the moment of starting. Thus, when the AC rotating magnetic field of the stator cuts the rotor windings, which are stationary at startup, the induced voltages produced may be high enough to damage the insulation if precautions are not taken. If the DC rotor field is either connected as a closed circuit or connected to a discharge resistor during the starting period, the resulting current produces a voltage drop that is opposed to the generated voltage. Thus, the induced voltage at the field terminal is reduced. The squirrel cage winding is used to start the synchronous motor in the same way it is used in the squirrel cage induction motor. When the rotor reaches the maximum speed to which it can be accelerated as a squirrel cage motor (about 95 percent or more of the synchronous speed), direct current is applied to the rotor field coils to establish north and south rotor poles. These poles then are attracted by the poles on the stator. The rotor then accelerates until it locks into synchronous motion with the stator field.

Synchronous motors are used for applications involving large, slow-speed machines with steady loads and constant speeds. Such applications include compressors, fans and pumps, many types of crushers and grinders, and pulp, paper, rubber, chemical, flour, and metal rolling mills.

Power Factor Correction by Synchronous Motor

A synchronous motor converts alternating current electrical energy into mechanical power. In addition, it also provides power factor correction. It can operate at a leading power factor or at unity. In very rare occasions, it can operate at a lagging power factor.

The *power factor* is of great concern to industrial users of electricity with respect to energy conservation. Power factor is the ratio of the actual power being used in a circuit, expressed in watts or kilowatts (WA), to the power apparently being drawn from the line, expressed in volt-amperes or kilovolt-amperes (kVA). The kVA value is obtained by multiplying a voltmeter reading and an ammeter reading of the same circuit or equipment. Inductance within the circuit will cause the current to lag the voltage.

When the values of the apparent and actual power are equal or in phase, the ratio of these values is 1:1. In other words, when the voltage and amperage are in phase, the ratio of these values is 1:1. This is the case of pure resistive loads.

The *unity* power factor value is the highest power factor that can be obtained. The higher the power factor, the greater the efficiency of the electrical equipment.

Alternating current loads generally have a lagging power factor. As a result, these loads burden the power system with a large reactive load.

A synchronous motor with an overexcited DC field may be used to offset the low power factor of the other loads on the same electrical system. An overexcited synchronous motor means that it is operating at more than the unity power factor. Therefore, it is working to improve the power factor of the power system.

Brushless Synchronous Motors

Solid-state technology has brought about the use of *brushless synchronous motors*. The DC field excitation for such a motor is provided by a special AC generator mounted on the main motor shaft. The excitation is converted to direct current by a rotating rectifier assembly.

The operating characteristics are the same as those of synchronous motors with brushes. However, elimination of the collector rings, brushes, commutator, and some control contactors gives the brushless motor several outstanding advantages:

- Brush sparking is eliminated, reducing safety hazards in some areas
- Field control and excitation are provided by a static system, requiring much less maintenance
- Field excitation is automatically removed whenever the motor is out of step

Automatic resynchronization can be achieved whenever it is practical.

Automatic Synchronous Motor Starter

An automatic synchronous motor starter can be used with a synchronous motor to provide automatic control of the startup sequence. That is, the controller automatically sequences the operation of the motor so that the rotor field is synchronized with the revolving magnetic field of the stator.

There are two basic methods of starting synchronous motors automatically. In the first method, full voltage is applied to the stator winding. In the second method, the starting voltage is reduced. A commonly used method of starting synchronous motors is the *across-the-line connection*. In this method, the stator of the synchronous motor is connected directly to the plant distribution system at full voltage. A magnetic starter is used in this method of starting.

A *polarized field frequency relay* can be used for the automatic application of field excitation to a synchronous motor.

Rotor Control Equipment

Field Contactor   The field contactor opens both lines to the source of excitation. During starting, the contactor also provides a closed field circuit through a discharge resistor. A solenoid-operated field contactor is similar in appearance to the standard DC contactor. However, for this DC operated contactor, the center pole is normally closed. It is designed to provide a positive overlap between the normally closed contact and the two normally open contacts. This overlap is an important feature because it means that the field winding is never open. The field winding of the motor must always be short-circuited through a discharge resistor or connected to the DC line. The coil of the field contactor is operated from the same direct current source that provides excitation for the synchronous motor field.

Out-of-Step Relay   The squirrel cage winding, or starting (amortisseur) winding will not overheat if a synchronous motor starts, accelerates, and reaches synchronous speed within a time interval determined to be normal for the motor. In addition, the motor must continue to operate at synchronous speed. Under these conditions, adequate protection for the entire motor is provided by three overload relays in the stator winding. The squirrel cage winding, however, is designed for starting only. If the motor operates at subsynchronous speed, the squirrel cage winding may overheat and be damaged. It is not unusual for some synchronous motors to withstand a maximum locked rotor interval of only five to seven seconds.

An *out-of-step relay* (OSR) is provided on automatic synchronous starters to protect the starting winding. The normally closed contacts of the relay will open to deenergize the line contactor under the following conditions:

1. The motor does not accelerate and reach the synchronizing point after a preset time delay.

2. The motor does not return to a synchronized state after leaving it.

3. The amount of current induced in the field winding exceeds a value determined by the core setting of the out-of-step relay.

As a result, power is removed from the stator circuit before the motor overheats.

Polarized Field Frequency Relay

A synchronous motor is started by accelerating the motor to as high a speed as possible from the squirrel cage winding and then applying the DC field excitation. The components responsible for correctly and dependably applying and removing the field excitation are a *polarized field frequency relay (PFR)* and a reactor.

Summary of Automatic Starter Operation

The line diagram in Figure 12–10 shows the automatic operation of a synchronous motor. For starting, the motor field winding is connected through the normally closed power contact of the field contactor (F), the discharge resistor, the coil of the out-of-step relay, and the reactor. When the start button is pressed, the circuit is completed to the control relay coil (CR_1) through the control fuses, the stop button, and contacts of the overload and out-of-step relays. The closing of CR_1 energizes the line contactor M which applies full voltage at the motor terminals with the overload relays in the circuit. A normally open contact on CR_1 and a normally open interlock on line contactor M provide the hold-in, or maintaining circuit. The starting and running current drawn by the motor is indicated by an ammeter with a current transformer.

At the moment the motor starts, the polarized field frequency relay (*PFR*) opens its normally closed contact and maintains an open circuit to the field contactor (F) until the motor accelerates to the proper speed for synchronizing. When the motor reaches a speed equal to 92 to 97 percent of its synchronous speed, and the rotor is in the correct position, the contact of the polarized field frequency relay closes to energize field contactor F through an interlock on line contactor M. The closing of field contactor F applies the DC excitation to the field winding and causes the motor to synchronize. After the rotor field circuit is established through the normally open power contacts of the field contactor, the normally closed contact on this contactor opens the discharge circuit. The motor is now operating at the synchronous speed. If the stop button is pressed, or if either magnetic overload relay is tripped, the starter is deenergized and disconnects the motor from the line.

Variable Speed Alternating Current Motor Control

Two factors determine the speed of the rotating magnetic field of an AC induction motor:

1. Number of stator poles.

2. Frequency.

If either of these factors is changed, the speed of the motor can be changed.

The speed of the consequent pole motor can be changed by changing the number of stator poles. This method of speed control causes the speed to change in steps and does not permit control over a wide range of speed. For example, the synchronous speed of the rotating magnetic field in a two-pole motor is 3600 rpm when connected to a 60 Hz line. If this motor is changed to a four-pole motor, the synchronous speed will change to 1800 rpm.

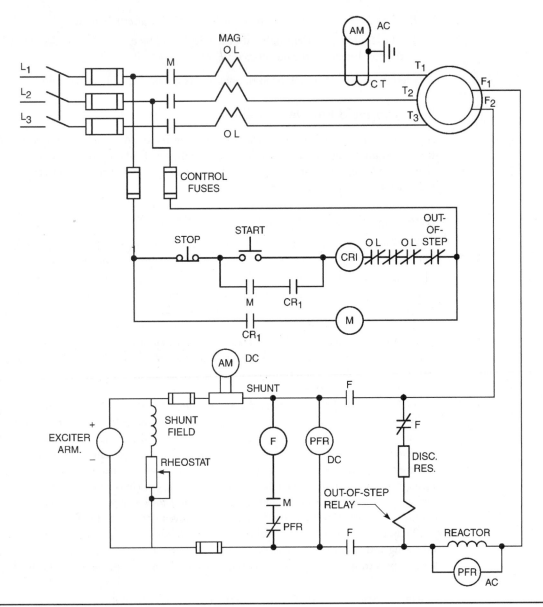

Figure 12–10 Line diagram for automatic operation of a synchronous motor using polarized field frequency relay.

Notice that this method of speed control permits the motor to operate with a synchronous field speed of 3600 rpm or 1800 rpm. The motor cannot be operated with a synchronous speed between 3600 rpm and 1800 rpm.

Variable Voltage Speed Control

Another method of controlling the speed of some AC induction motors is by reducing the applied voltage to the stator. This method does not change the synchronous speed of the

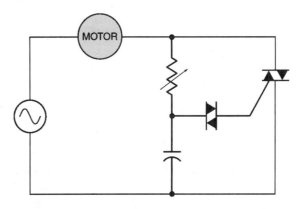

Figure 12–11   Triac used to control motor speed.

rotating magnetic field, but it does cause the magnetic field of the stator to become weaker. As the magnetic field of the stator becomes weaker, the rotor slip becomes greater, thereby causing a reduction in rotor speed.

Variable voltage speed control is used with fractional horsepower motors that operate light loads such as fans or blowers. Motors that are intended to be operated with variable voltage are designed with high resistance rotors, such as the type A rotor, to help limit the amount of current induced into the rotor at low speed. Induction motors that use a centrifugal switch cannot be used with variable voltage control. This limits the types of induction motors that can be used to shaded pole motors or capacitor-start/capacitor-run motors.

There are several methods used to control the voltage supplied to the motor. One method is to use a *triac* with a phase shift network similar to the circuit shown in Figure 12–11. When triac circuits are used with inductive loads, care must be taken to insure that both halves of the AC waveform are conducted. Only triac controllers intended to be used with inductive loads should be used for motor control. Triac circuits intended to control incandescent lamp loads will often begin conducting on one half of the AC cycle and not the other. For example, assume the triac in Figure 12–11 begins conducting on the positive half cycle of voltage before it conducts the negative half cycle. Since only positive voltage pulses are being conducted, the voltage applied to the load is DC. A DC voltage applied to a resistive load such as an incandescent lamp will not cause any harm, but a great deal of harm can be done if a DC voltage is applied to an inductive load such as the stator winding of a motor.

Another device used to control the voltage applied to a small induction motor is the *autotransformer*. In this circuit, a rotary switch is used to connect the motor to different taps on the transformer winding. This permits the motor to be operated at any one of several different speeds.

Variable Frequency Control

One of the factors that determines the speed of the rotating magnetic field of an induction motor is the frequency of the applied voltage. If the frequency is changed, the speed of the rotating magnetic field also changes. For example, a four-pole stator connected to a 60 Hz line will have a synchronous speed of 1800 rpm. If the frequency is lowered to 30 Hz, the synchronous field speed falls to 900 rpm.

When the frequency is lowered, care must be taken not to damage the stator winding. The current flow through the winding is limited to a great extent by inductive reactance. When the frequency is lowered, inductive reactance is lowered also ($XL = 2\pi FL$). For this reason, variable frequency motor controllers must have some method of reducing the applied voltage to the stator as frequency is reduced.

Alternator Control

One method of producing variable frequencies for operating induction motors is with the use of an *alternator*. In this arrangement, some type of variable speed drive, such as DC motor, is used to turn the shaft of the alternator. The speed of the alternator determines the frequency of the voltage applied to the induction motors. The alternator can furnish power to as many induction motors as desired providing the power rating of the alternator is not exceeded.

Since the output voltage of the alternator is controlled by the amount of DC excitation current applied to the rotor, a variable voltage DC power supply is used to determine the output voltage of the alternator. As frequency is reduced, the output voltage must also be reduced to prevent excessive current flow in the windings of the induction motors. This method of speed control is frequently used on conveyor systems where it is desirable to have a large number of motors controlled from one source.

Solid-State Control

Most solid-state variable frequency drives operate by first changing the AC voltage to DC, and then changing the DC voltage back to AC at the desired frequency. The circuit shown in Figure 12–12 uses a three-phase bridge rectifier to convert three-phase AC voltage into DC voltage. A phase shift unit controls the output voltage of the rectifier. This permits the voltage applied to the motor to be decreased as the frequency is decreased.

A *choke coil* and *capacitor bank* are used to filter the output voltage of the rectifier before transistors $Q1$ through $Q6$ change the DC voltage back to AC. An electronic control unit is connected to the bases of transistors $Q1$ through $Q6$. The electronic control unit converts the DC voltage back into three-phase alternating current by turning transistors on or off at the proper time and in the proper sequence. For example, assume that transistors $Q1$ and $Q4$ are switched on at the same time. This permits T_1 to be positive at the same time T_2 is negative. If conventional current flow is assumed, current will flow through transistor Q_1 to T_1, from T_1 through the motor winding to T_2, and then through transistor Q_4 to negative. Now assume that transistors Q_1 and Q_4 have been turned off, and transistors Q_3 and Q_6 have been turned on. Current can now flow through transistor Q_3 to T_2, from T_2 through the motor to T_3, and through transistor Q_6 to negative.

Since the transistors are turned either completely on or completely off, the waveform produced is a square wave instead of a sine wave. Induction motors will operate on a square waveform without any problems. Some manufacturers design units that will produce a *stepped waveform*. This stepped waveform is used because it is similar to a sine wave.

Variable Frequency Control Using Silicon-Controlled Rectifiers

Because of their ability to handle large amounts of power, silicon-controlled rectifiers (SCRs) are often used for converting direct current into alternating current. An example of this type of circuit is illustrated in Figure 12–13. In this circuit, the SCRs are connected to a control unit which controls the sequence and rate at which the SCRs are gated "on." The

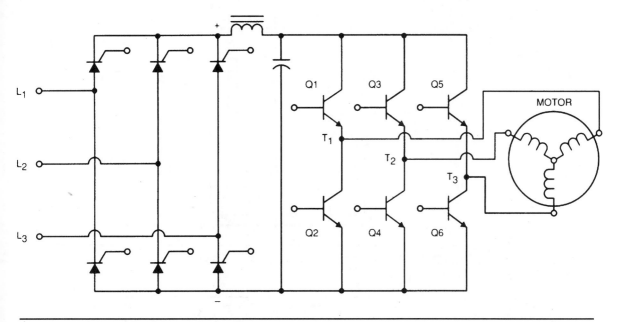

Figure 12–12   Solid-state variable frequency control.

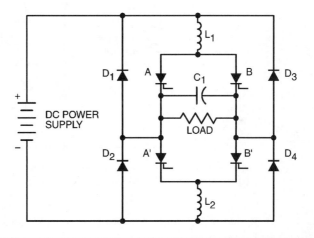

Figure 12–13   Changing DC into AC using SCRs.

circuit is constructed so that SCRs A and A' and SCRs B and B' are simultaneously gated "on." Inductors L_1 and L_2 are used for filtering and wave shaping. Diodes D through $D4$ are clamping diodes and are used to prevent the output voltage from becoming excessive. Capacitor C is used to turn one set of SCRs off when the other set is gated on. This capacitor must be a true AC capacitor because it will be charged to the alternate polarity each half cycle. In a converter intended to handle large amounts of power, capacitor C will be a bank of capacitors. To understand the operation of this circuit, assume that SCRs A and A' are gated on at the same time. Current will flow through the circuit as shown in Figure 12–14. Notice the direction of current flow through the load and that capacitor C has been charged

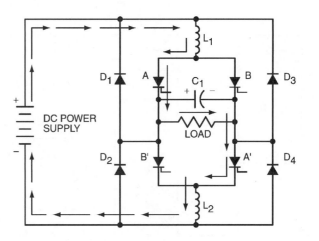

Figure 12–14 Current flows through SCRs A and A'.

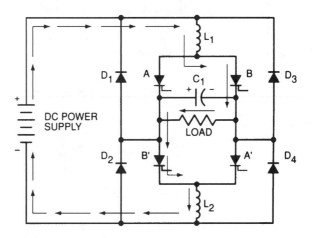

Figure 12–15 Current flows through SCRs B and B'.

to the polarity shown. Recall that when an SCR has been turned on by the gate, it can only be turned off by permitting the current flow through the anode cathode section to drop below the holding current level. Now assume that SCRs *B* and *B'* are gated on. Because SCRs *A* and *A'* are still turned on, two separate current paths now exist through the circuit. The negative charge on capacitor *C*, however, causes the positive current to see a path more negative than the one through SCRs *A* and *A'*. The current now flows through SCRs *B* and *B'* to charge capacitor *C*, to the opposite polarity as shown in Figure 12–14. Because the current now flows through SCRs *B* and *B'*, SCRs *A* and *A'* turn off. Notice that the current flows through the load in the opposite direction, which produces alternating current through the load, and that capacitor *C* has been charged to the opposite polarity.

To produce the next half cycle of AC current, SCRs *A* and *A'* are gated on again. The negatively charged side of capacitor *C* will now cause the current to stop flowing through SCRs *B* and *B'* and begin flowing through SCRs *A* and *A'* as shown in Figure 12–15. The frequency of the circuit is determined by the rate at which the SCRs are gated on.

Drives

The Purpose and Types of Drives

Controlling speed of the motor is the primary purpose of a *drive*. Sometimes a drive is referred to as an *inverter*. This will be explained later in this text. There are several types of drives available for use.

Mechanical (Mechanical drives are usually pulley- or belt-type drives that change the speed mechanically.)
Hydraulic
Eddy Current
Rotating DC
Solid-State DC
Solid-State AC

Electrical Components of a Motor

Figure 13–1 represents a simplified equivalent circuit of an AC motor. An understanding of this diagram is important in the understanding of how an AC motor is applied to an AC drive.

V_S	Line voltage applied to stator power leads
R_S	Stator resistance
L_S	Stator leakage inductance
E	Air gap or magnetizing voltage
L_m	Magnetizing inductance
I_m	Magnetizing current
R_R	Rotor resistance (varies with temperature)
L_R	Rotor leakage inductance
I_w	Working or torque producing current
I_S	Stator current

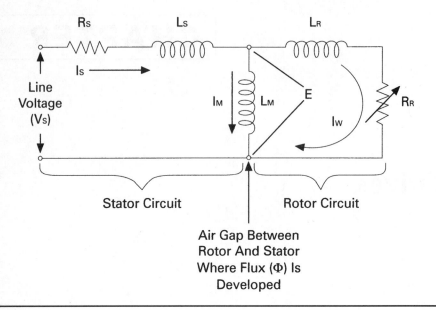

Figure 13–1 Electrical components of a motor. Courtesy Siemens Energy and Automation, Inc.

Line Voltage

Voltage (*Vs*) is applied to the stator power leads from the AC power supply. Voltage drops occur due to stator resistance (*Rs*). The resultant voltage (*E*) represents counter-electromagnetic force (cemf) available to produce magnetizing flux and torque.

Magnetizing Current

Magnetizing current (Im) is responsible for producing magnetic lines of flux that magnetically link with the rotor circuit. Magnetizing current is typically about 30% of rated current. Magnetizing current, like flux (Φ), is proportional to voltage (*E*) and frequency (*F*).

$$Im = \frac{E}{2 \ \pi FL \ m}$$

Working Current

The current that flows in the rotor circuit and produces torque is referred to as *working current (Iw)*. Working current is a function of the load. An increase in load causes the rotor circuit to work harder, increasing working current. A decrease in load decreases the work the rotor circuit does, decreasing working current.

Stator Current

Stator current (Is) is the current that flows in the stator circuit. Stator current can be measured on the supply line and is also referred to as line current. A clamp-on ammeter, for example, is frequently used to measure stator current. The full-load ampere rating on the nameplate of a motor refers to stator current at rated voltage, frequency, and load. It is the maximum current the motor can carry without damage. Stator current is the vector sum of working current (*Iw*) and magnetizing current (*Im*). Typically, magnetizing current (*Im*) remains constant. Working current will vary with the applied load which causes a corresponding change in stator current (*Is*).

$$Is = \sqrt{Im} + Iw \text{ squared}$$

The National Electrical Manufacturers Association (NEMA) classifies motors according to locked rotor torque and current, pull up torque, breakdown torque, and percent slip. In addition, full-load torque and current must be considered when evaluating an application.*

Definitions

Locked rotor torque   Also referred to as *starting torque*. This torque is developed when the rotor is held at rest with rated voltage and frequency applied. This condition occurs each time a motor is started. When rated voltage and frequency are applied to the stator there is a brief amount of time before the rotor turns.

Locked rotor current   Also referred to as *starting current*. This is the current taken from the supply line at rated voltage and frequency with the rotor at rest.

Pull up torque   The torque developed during acceleration from start to the point breakdown torque occurs.

Breakdown torque   The maximum torque a motor develops at rated voltage and speed without an abrupt loss of speed.

Full-load torque   The torque developed when the motor is operating with rated voltage, frequency, and load.

Full-load current   The current taken from the supply line at rated voltage, frequency, and load.

NEMA Classifications

Three-phase AC motors are classified by NEMA as NEMA A, B, C, and D. NEMA specifies certain operating characteristics for motors when started by applying rated voltage and frequency (across the line starting). A NEMA B motor, for example, typically requires 600 percent starting current and 150 percent starting torque. These considerations do not apply to motors started with an AC drive. NEMA B design motors are the most common and most suitable for use on AC drives.

NEMA B Speed/Torque Curve

The graph in Figure 13–2 illustrates the relationship between motor speed and torque of a NEMA B motor. When rated voltage and frequency are applied to the motor, synchronous speed goes to 100 percent immediately. The rotor must perform a certain amount of work to overcome the mechanical inertia of itself and the connected load. Typically a NEMA B motor will develop 150 percent torque to start the rotor and load. As the rotor accelerates, the relative difference in speed between synchronous speed and rotor speed decreases until the rotor reaches its operating speed. The operating speed of a NEMA B motor with rated voltage, frequency and load is approximately 97 percent (3 percent slip) of synchronous speed. The amount of slip and torque is a function of load. With an increase in load there

*Most NEMA terms and concepts apply to motors operated from 60 Hz power lines, not variable speed drive operation.

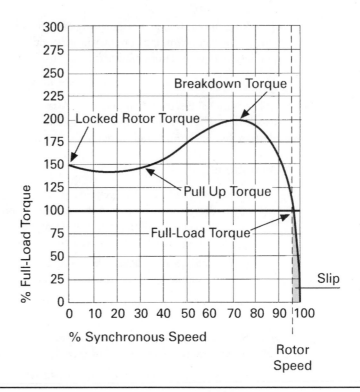

Figure 13–2 Speed and torque graph of a NEMA B motor. Courtesy Siemens Energy and Automation, Inc.

is a corresponding increase in slip and torque. With a decrease in load there is a corresponding decrease in slip and torque.

Starting Current

When a motor is started, it must perform work to overcome the inertia of the rotor and attached load. The starting current measured on the incoming line (Is) is typically 600 percent of full-load current when rated voltage and frequency is first applied to a NEMA B motor. Stator current decreases to its rated value as the rotor comes up to speed. The following graph applies to "across the line" operation, not variable speed drive operation.

Voltage and Frequency

Up to this point, the operation of an AC motor with rated voltage and frequency applied has been examined. Many applications require the speed of an AC motor to vary. The easiest way to vary the speed of an AC induction motor is to use an AC drive to vary the applied frequency. Operating a motor at other than the rated frequency has an effect on motor current and torque.

Volts per Hertz

A ratio exists between voltage and frequency. This ratio is referred to as *volts per hertz* (*V/Hz*). A typical AC motor manufactured for use in the United States is rated for 460 VAC and 60 Hz. The ratio is 7.67 volts per hertz, as illustrated by the equation:

$$\frac{460}{60} = 7.67 \ V/Hz$$

Not every motor has a 7.67 V/Hz ratio. A 230-Volt, 60-Hz motor, for example, has a 3.8 V/Hz ratio:

$$\frac{230}{60} = 3.8 \ V/Hz$$

Flux (Φ), magnetizing current (Im), and torque are all dependent on this ratio. Increasing frequency (F) without increasing voltage (E), for example, will cause a corresponding increase in speed. However, flux will decrease, causing motor torque to decrease. Magnetizing current (Im) will also decrease. A decrease in magnetizing current will cause a corresponding decrease in stator or line (Is) current. These decreases are all related and greatly affect the motor's ability to handle a given load.

$$\Phi = \frac{E}{F}$$
$$T = k\Phi Iw$$
$$Im = \frac{E}{2\pi \ FLm}$$

Constant Torque

Alternating current motors running on an AC line operate with a constant flux (Φ) because voltage and frequency are constant. Motors operated with constant flux are said to have *constant torque*. Actual torque produced, however, is determined by the demand of the load.

$$T = k \ \Phi \ Iw$$

An AC drive is capable of operating a motor with constant flux from approximately zero (0) to the motor's rated nameplate frequency (typically 60 Hz). This is the constant torque range. As long as a constant volts per hertz ratio is maintained, the motor will have constant torque characteristics. AC drives change frequency to vary the speed of a motor and voltage proportionately to maintain constant flux. The following graphs illustrate the volts per hertz ratio of a 460 volt, 60 Hz motor and a 230 volt, 60 Hz motor. To operate the 460 volt motor at 50 percent speed with the correct ratio, the applied voltage and frequency would be 230 volts, 30 Hz. To operate the 230 volt motor at 50 percent speed with the correct ratio, the applied voltage and frequency would be 115 volts, 30 Hz. The voltage and frequency ratio can be maintained for any speed up to 60 Hz. This usually defines the upper limits of the constant torque range.

Constant Horsepower

Some applications require the motor to be operated above base speed. The nature of these applications requires less torque at higher speeds. Voltage, however, cannot be higher than the rated nameplate voltage. This can be illustrated using a 460 volt, 60 Hz motor. Voltage will remain at 460 volts for any speed above 60 Hz. A motor operated above its rated frequency is operating in a region known as a *constant horsepower*. Constant volts per hertz and torque is maintained to 60 Hz. Above 60 Hz the volts per hertz ratio decreases.

Frequency	V/Hz
30 Hz	7.67
60 Hz	7.67
70 Hz	6.6
90 Hz	5.1

Flux (Φ) and torque (T) decrease:

$$\Phi = \frac{E}{F} \qquad T = \frac{k\Phi}{w}$$

Horsepower remains constant as speed (N) increases and torque decreases in proportion. The following formula applies to speed in revolutions per minute (RPM).

$$HP \; (remains \; consistant) = \frac{T \; (decreases) \times N \; (increases)}{5250}$$

Reduced Voltage and Frequency Starting

A NEMA B motor that is started by connecting it to the power supply at full voltage and full frequency will develop approximately 150 percent starting torque and 600 percent starting current. Alternating current drives start at reduced voltage and frequency. The motor will start with approximately 150 percent torque and 150 percent current at reduced frequency and voltage. The torque/speed curve shifts to the right as frequency and voltage are increased. The dotted lines on the torque/speed curve illustrated below represent the portion of the curve not used by the drive. The drive starts and accelerates the motor smoothly as frequency and voltage are gradually increased to the desired speed. An AC drive, properly sized to a motor, is capable of delivering 150 percent torque at any speed up to speed corresponding to the incoming line voltage. The only limitations on starting torque are peak drive current and peak motor torque, whichever is less.

Some applications require higher than 150 percent starting torque. A conveyor, for example, may require 200 percent rated torque for starting. If a motor is capable of 200 percent torque at 200 percent current, and the drive is capable of 200 percent current, then 200 percent motor torque is possible. Typically drives are capable of producing 150 percent of drive nameplate rated current for one minute. A drive with a larger current rating would be required. It is appropriate to supply a drive with a higher continuous horsepower rating than the motor when high peak torque is required.

Selecting a Motor

Alternating current drives often have more capability than the motor. Drives can run at higher frequencies than may be suitable for an application. In addition, drives can run at low speeds. Self-cooled motors may not develop enough air flow for cooling at reduced speeds and full load. Consideration must be given to the motor.

The following graph indicates the speed and torque range of a sample motor. Each motor must be evaluated according to its own capability. The sample motor can be operated continuously at 100 percent torque up to 60 Hz. Above 60 Hz the V/Hz ratio decreases and the motor cannot develop 100 percent torque. This motor can be operated continuously at 50 percent torque at 120 Hz. The motor is also capable of operating above rated torque intermittently. The motor can develop as much as 150 percent* torque for

*Torque may be higher than 150 percent if the drive is capable of higher current.

starting, accelerating, or load transients, if the drive can supply the current. At 120 Hz the motor can develop 70 percent torque intermittently.

The sample motor described above is capable of operating at 100 percent rated torque continuously at low frequencies. Many motors are capable of operating continuously at 100 percent continuous torque at low frequencies. Each motor must be evaluated before selecting it for use on an AC drive.

Basic Alternating Current Drives

Alternating current drives, inverters, and adjustable frequency drives are all terms that are used to refer to equipment designed to control the speed of an AC motor. The AC drives receive AC power and convert it to an adjustable frequency, adjustable voltage output for controlling motor operation. A typical inverter receives 480 VAC, three-phase, 60 Hz input power and provides power to the motor which can be steplessly adjusted through the speed range. The three common inverter types are (1) the *variable voltage inverter* (*VVI*), (2) *current source inverter* (*CSI*), and (3) *pulse width modulation* (*PWM*). All AC drives convert AC to DC, and then through various switching techniques invert the DC into a variable voltage, variable frequency output.

Variable Voltage Inverter (VVI)

The variable voltage inverter uses an SCR converter bridge to convert the incoming AC voltage into DC. The SCRs provide a means of controlling the value of the rectified DC voltage from 0 to approximately 600 VDC. The L_1 choke and C_1 capacitor(s) make up the DC link section and smooth the converted DC voltage. The inverter section consists of six switching devices. Various devices can be used such as thyristors, bipolar transistors, MOSFETS, and IGBTs. The following schematic shows an inverter that utilizes bipolar transistors. Control logic (not shown) uses a microprocessor to switch the transistors on and off, providing a variable voltage and frequency to the motor. (See Figure 13–3.)

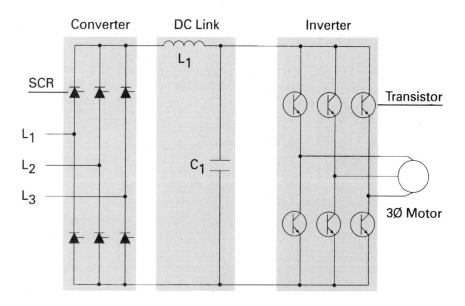

Figure 13–3   Variable voltage inverter (VVI). Courtesy Siemens Energy and Automation, Inc.

This type of switching is often referred to as six-step because it takes six 60-degree steps to complete one 360-degree cycle. Although the motor prefers a smooth sine wave, a six-step output can be satisfactorily used. The main disadvantage is torque pulsation which occurs each time a switching device, such as a bipolar transistor, is switched. The pulsations can be noticeable at low speeds as speed variations in the motor. These speed variations are sometimes referred to as cogging. The non-sinusoidal current waveform causes extra heating in the motor requiring a motor derating.

Current Source Inverter (CSI)

The current source inverter uses an SCR input to produce a variable voltage DC link. The inverter section also uses SCRs for switching the output to the motor. The current source inverter controls the current in the motor. The motor must be carefully matched to the drive. (See Figure 13–4.) Current spikes, caused by switching, can be seen in the output. At low speeds current pulses can causes the motor to cog.

Pulse Width Modulation (PWM)

Pulse width modulation drives provide a more sinusodial current output to control frequency and voltage supplied to an AC motor. PWM drives are more efficient and typically provide higher levels of performance. A basic PWM drive consists of a converter, DC link, control logic, and an inverter. (See Figure 13–5.)

Converter and DC Link

The converter section consists of a fixed diode bridge rectifier which converts the three-phase power supply to a DC voltage. The L_I choke and C_I capacitor(s) smooth the converted DC voltage. The rectified DC value is approximately 1.35 times the line-to-line

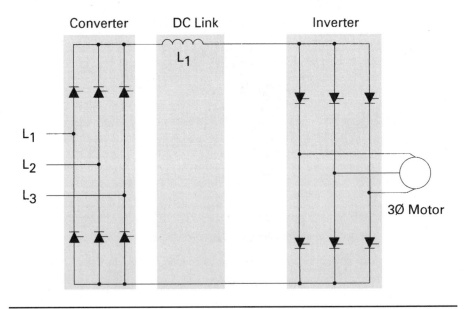

Figure 13–4 Current source inverter (CSI). Courtesy Siemens Energy and Automation, Inc.

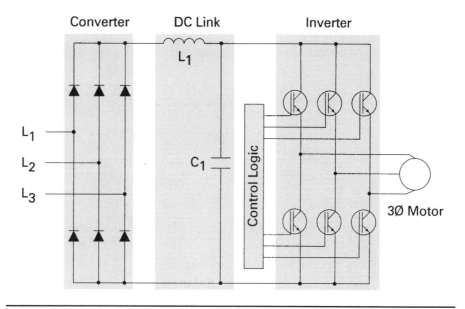

Figure 13–5 Pulse width modulation (PWM). Courtesy Siemens Energy and Automation Inc.

value of the supply voltage. The rectified DC value is approximately 650 VDC for a 480 VAC supply (Figure 13–6).

Control Logic and Inverter

Output voltage and frequency to the motor are controlled by the control logic and inverter section. The inverter section consists of six switching devices. Various devices can be used such as thyristors, bipolar transistors, MOSFETS, and insulated gate bipolar transistors (IGBTs). The following schematic shows an inverter that utilizes IGBTs. The control logic uses a microprocessor to switch the IGBTs on and off, providing a variable voltage and frequency to the motor. (See Figure 13–7.)

Isolated Gate Bipolar Transistors

Isolated gate bipolar transistors (IGBTs) provide a high switching speed necessary for PWM inverter operation. Isolated gate bipolar transistors are capable of switching on and off several thousand times a second. An IGBT can turn on in less than 400 nanoseconds and off in approximately 500 nanoseconds. An IGBT consists of a gate, collector, and an emitter. When a positive voltage (typically +15 VDQ) is applied to the gate, the IGBT will turn on. This is similar to closing a switch. Current will flow between the collector and emitter. An IGBT is turned off by removing the positive voltage from the gate. During the off state the IGBT gate voltage is normally held at a small negative voltage (–15 VDQ) to prevent the device from turning on (see Figure 13–8).

Using Switching Devices to Develop a Basic Alternating Current Output

In the following example, one phase of a three-phase output is used to show how an AC voltage can be developed. Switches replace the IGBTs. A voltage that alternates between positive and negative is developed by opening and closing switches in a specific sequence.

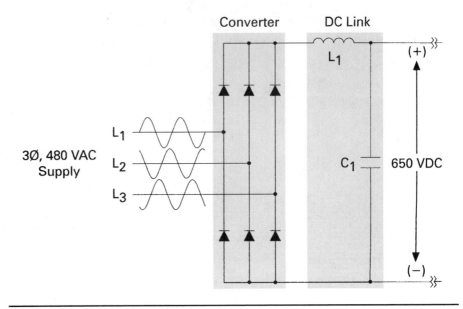

Figure 13-6 Converter and DC link. Courtesy Siemens Energy and Automation, Inc.

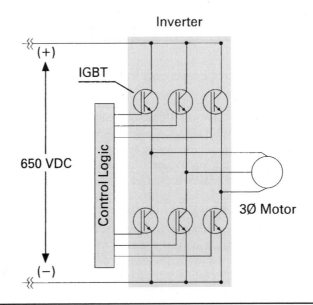

Figure 13-7 Control logic and inverter. Courtesy Siemens Energy and Automation, Inc.

For example, during steps one and two A+ and B– are closed. The output voltage between A and B is positive. During step three A+ and B+ are closed. The difference of potential from A to B is zero. The output voltage is zero. During step four A– and B+ are closed. The output voltage from A to B is negative. The voltage is dependent on the value of the DC voltage and the frequency is dependent on the speed of the switching. An AC sine wave has been added to the output (A – B) to show how AC is simulated (Figure 13–9).

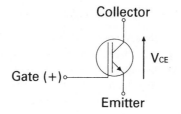

Figure 13–8   Isolated gate bipolar transistor (IGBT). Courtesy Siemens Energy and Automation, Inc.

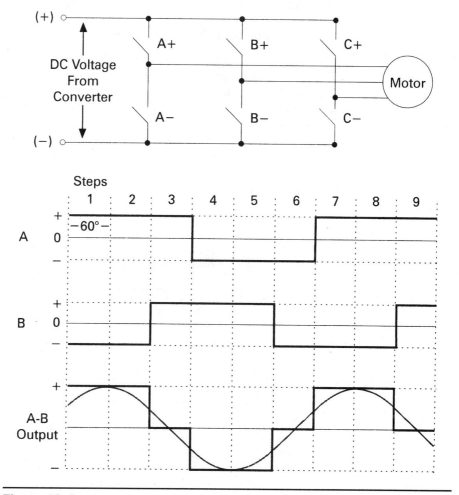

Figure 13–9   Using switching devices to develop a basic AC output. Courtesy Siemens Energy and Automation, Inc.

Generating a PWM Output

There are several PWM modulation techniques. It is beyond the scope of this book to describe all of them in detail. The following text and illustrations describe a typical pulse width modulation method. An IGBT (or other type switching device) can be switched on connecting the motor to the positive value of DC voltage (650 VDC from the converter). Current flows in the motor. The IGBT is switched on for a short period of time, allowing only a small amount of current to build up in the motor and then switched off. The IGBT is switched on and left on for progressively longer periods of time, allowing current to build up to higher levels until current in the motor reaches a peak. The IGBT is then switched on for progressively shorter periods of time, decreasing current build up in the motor. The negative half of the sine wave is generated by switching an IGBT connected to the negative value of the converted DC voltage (See Figure 13–10.)

PWM Voltage and Current Waveforms

The more sinusoidal current output produced by the PWM reduces the torque pulsations, low speed motor cogging, and motor losses noticeable when using a six-step output.

The voltage and frequency is controlled electronically by circuitry within the AC drive. The fixed DC voltage (650 VDC) is modulated or clipped with this method to provide a variable voltage and frequency. At low output frequencies a low output voltage is required. The switching devices are turned on for shorter periods of time. Voltage and current buildup in the motor is low. At high output frequencies a high voltage is required. The switching devices are turned on for longer periods of time. Voltage and current buildup in the motor increases. (See Figure 13–11.)

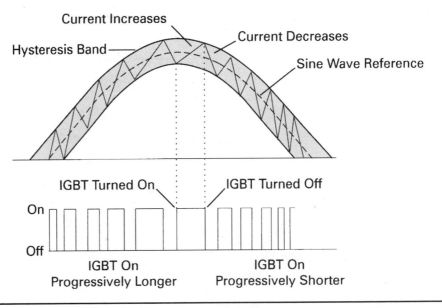

Figure 13–10 Generating a pulse width modulation output. Courtesy Siemens Energy and Automation, Inc.

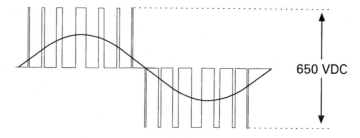

Shorter "On" Duration, Lower Voltage

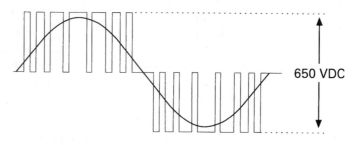

Longer "On" Duration, Higher Voltage

Figure 13–11 Pulse width modulation wave forms. Courtesy Siemens Energy and Automation, Inc.

Distance Between Drive and Motor

Distance from the drive to the motor must also be taken into consideration. All motor cables have line-to-line and line-to-ground capacitance. The longer the cable, the greater the capacitance. Some types of cables (shielded cable, for example) have greater capacitance. Spikes occur on the output of all PWM drives because of the charging current of the cable capacitance. Higher voltage (460 VAC and higher capacitance long cables) result in higher current spikes. Voltage spikes caused by long cable lengths can potentially shorten the life of the inverter and the motor. So when considering an application, consider if distance may be a problem. The maximum distance between a motor and the drive is 50 meters (64 feet). If shielded cable is used, the maximum distance is 25 meters (82 feet).

230 Volt Drives

Input voltage is single-phase 208 to 230 VAC. In addition, there is a 10 percent /+ 15 percent (187 VAC to 264 VAQ) tolerance for voltage fluxuations. The acceptable input frequency ranges from 47 Hz to 63 Hz. The output voltage is 0 to 200/230 VAC, three-phase. The maximum value depends on the input voltage.

380 to 500 Volt Drives

Input voltage is three-phase 380 to 500 VAC. In addition, there is a 10 percent (342 VAQ) to + 10 percent (550 VAC) tolerance for voltage fluxuations. The acceptable input frequency ranges from 47 Hz to 63 Hz. The output voltage is 0 to 400/500 VAC, three-phase. The maximum value depends on the input voltage.

Output Frequency

Output frequency is adjustable from 0 to 650 Hz on most drives. Consideration must be given to the motor and application, which may not be capable of operating at 650 Hz. The drive is capable of being set to a minimum and maximum speed appropriate for the application. If the minimum speed required is 10 Hz, and the maximum speed required is 60 Hz, then the drive can be programmed for a speed range of 10 to 60 Hz.

Parameters

A *parameter* is a variable that is given a constant value. For example, ramp up time is parameter number P002. Ramp up time can be set for any value from 0 to 650 seconds. When setting up the drive, a value of 10 seconds might be selected. When given a start command it would take 10 seconds for the drive to ramp up to full speed. While standard applications require no special settings, parameters can be easily modified for special applications.

Parameterization Panel

Programming is done on some drives using three push buttons and a 4-digit LED display. Parameters, such as ramp times, minimum and maximum frequencies, and modes of operation are easily set. The "P" key toggles the display between a parameter number and the value of the parameter. The up and down pushbuttons scroll through parameters and are used to set a parameter value. In the event of a failure, the inverter switches off and a fault code appears in the display.

Drives use various parameters for proper drive and motor operation. It is beyond the scope of this book to describe in detail every parameter available. Some of the key parameters are described in the following text.

Ramp Function

A feature of AC drives is the ability to increase or decrease the voltage and frequency to a motor gradually. This accelerates the motor smoothly, with less stress on the motor and connected load. Parameters P002, P003, and P004 are used to set a ramp function. Acceleration and deceleration are separately programmable from 0 to 650 seconds. Acceleration, for example, could be set for 10 seconds and deceleration could be set for 60 seconds.

Analog Inputs

Parameter P006 is used to select either an analog or digital reference, or fixed frequency. Terminals one (1) through four (4) are used to provide an analog reference that controls the speed of the motor from 0 to 100 percent. Terminal one (1) is a +10 VIDC power supply that is internal to the drive. Terminal two (2) is the return path, or ground, for the 10 volt supply. An adjustable resistor can be connected between terminals one and two. Terminal three (3) is the positive (+) analog input to the drive. A positive voltage will control the speed of the motor in the forward direction. The wiper arm of the adjustable resistor is connected to terminal three. Terminal four (4) is a negative (−) analog input. A negative voltage will control the speed of the motor in the reverse direction. The drive can also be programmed to accept 0 to 20 mA, or 4 to 20 mA speed reference signal. These signals are typically supplied to the drive by other equipment such as a programmable logic controller (PLC). Note that a jumper has been connected between terminals two and four. An analog input cannot be left floating (open). If an analog input will not be used it must be connected to termini two (0 Volts).

Thermistor

Some motors have a built in *thermistor*. If a motor becomes overheated the thermistor acts to interrupt the power supply to the motor. A thermistor can be connected to terminals five (5) and six (6). If the motor gets to a preset temperature as measured by the thermistor, the drive will interrupt power to the motor. The motor will coast to a stop. The display will indicate a fault has occurred. Parameter P087 will tell the drive if a thermistor is used.

Digital Inputs

An input that is either "high" or "low" is referred to as a *digital input* (*DIN*). The drive has five digital inputs (DIN1-DIN5). A switch or a contact is connected between + 15 VDC (P1 5+) and a digital input. Digital inputs are controlled by parameters P051 through P052 and can be programmed for various functions. Digital input 1 (DIN1), for example, may be used to start and stop the drive. When the switch between P15+ is open the drive is stopped. When the switch between P15+ is closed the drive supplies power to the motor. Other inputs can be used for forward/reverse, preset speeds, jogging, or other functions.

Relay Outputs

There are two programmable relay outputs for indication of system status. These are controlled by parameters P061 and P062. One output (terminals 16, 17, 18) can be used for one normally closed (NC) or one normally open (NO) output. The other output (terminals 19, 20) can be used for a normally open output. The relays can be programmed to indicate various conditions, such as the drive is running, a failure has occurred, converter frequency is at 0, or converter frequency is at minimum. (See Figure 13–12.)

Serial Interface

Some drives have a RS485 serial interface that allows communication with computers (PCs) or programmable logic controllers (PLCs). The bus address is entered with parameter P091. Other parameters necessary for proper RS485 operation are P092-PO94. Several AC drives

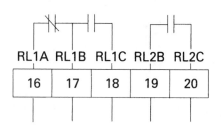

Figure 13–12 Relay output terminals. Courtesy Siemens Energy and Automation, Inc.

can be interconnected for control of multi-drive applications. The standard RS485 protocol is called USS protocol and is programmable up to 19.2 K baud (see Figure 13–13).

Current Limit and Motor Overload

Some drives are capable of delivering up to 150 percent of drive-rated current for one minute. Some drives also have a sophisticated and useful speed/time/current dependent overload function to protect the motor. The monitoring and protection functions include a drive overcurrent fault, a motor overload fault, a calculated motor overtemperature warning, and a measured motor overtemperature fault (requires a device inside the motor).

Low Speed and Starting Boost

Low speed boost can be adjusted high for applications requiring high torque at low speeds. Low speed boost can be adjusted low for smooth, cool, and quiet operation at low speed. An additional starting boost is available for applications requiring high starting torque.

Modes of Operation

There are three modes of operation, selected by parameter P077. The drive can operate utilizing a standard V/Hz curve. Using a 460 VAC, 60 Hz motor as an example, constant volts per hertz is supplied to the motor at any frequency between 0 and 60 Hz. This is the simplest type of control and is suitable for general purpose applications.

A second mode of operation is referred to as a quadratic voltage/frequency curve. This mode provides a V/Hz curve that matches the torque requirements of simple fan and pump applications. The third mode of operation is flux current control and warrants further discussion.

Flux Current Control (FCC)

Flux current control is the most complex of operation modes. Stator current (*Is*) is made up of active and reactive current. The reactive current component of stator current produces the rotating magnetic field. The active current produces work. Motor nameplate data is entered into the drive. The drive estimates motor magnetic flux based on the measured reactive stator current and the entered nameplate data. Proprietary internal computer algorithms attempt to keep the estimated magnetic flux constant.

If the motor nameplate information has been correctly entered and the drive properly set up, the flux current control mode will usually provide better dynamic performance than simple V/Hz control. Flux current control automatically adapts the drive output to the

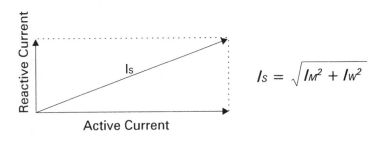

$$Is = \sqrt{I_M{}^2 + I_W{}^2}$$

Figure 13–13 Speed and current relationship. Courtesy Siemens Energy and Automation, Inc.

load. The motor is always operated at optimum efficiency. Speed remains reliably constant even under varying load conditions. (See Figure 13–13.)

Single-Quadrant Operation

Figure 13–14 illustrates that there are four quadrants according to direction of rotation and direction of torque. A *single-quadrant* drive operates only in quadrants I or III. Quadrant I is forward motoring or driving (*CW*). Quadrant III is reverse motoring or driving (*CCW*). Reverse motoring is achieved by reversing the direction of the rotating magnetic field. Motor torque is developed in the positive direction to drive the connected load at a desired speed (*N*). This is similar to driving a car forward on a flat surface from standstill to a desired speed. It takes more forward or motoring torque to accelerate the car from zero to the desired speed. Once the car has reached the desired speed, your foot can be let off the accelerator a little. When the car comes to an incline, a little more gas, controlled by the accelerator, maintains speed.

To stop an AC motor in single-quadrant operation, voltage and frequency can simply be removed and the motor allowed to coast to a stop. This is similar to putting a car in neutral, turning off the ignition, and allowing the car to coast to a stop. Another way is to use a controlled deceleration. Voltage and frequency are reduced gradually until the motor is at stop. This would be similar to slowly removing your foot from the accelerator of a car. The amount of time required to stop a motor depends on the inertia of the motor and connected load. The more inertia, the longer it will take to stop.

Direct Current Injection Braking

The braking mode stops the rotating magnetic field and applies DC voltage to the motor windings, helping stop the motor. Up to 250 percent of the motor's rated current can be

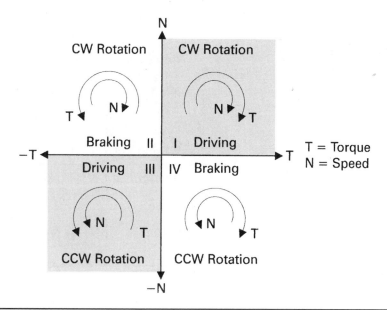

Figure 13–14 Single-quadrant operation. Courtesy Siemens Energy and Automation, Inc.

applied. This is similar to removing your foot from the accelerator and applying the brakes to bring the car to a stop quickly.

Four-Quadrant Operation

The dynamics of certain loads may require *four-quadrant* operation. When equipped, an optional braking resistor is capable of four-quadrant operation. Torque will always act to cause the rotor to run towards synchronous speed. If the synchronous speed is suddenly reduced, negative torque is developed in the motor. The motor acts like a generator by converting mechanical power from the shaft into electrical power which is returned to the AC drive. This is similar to driving a car downhill. The car's engine will act as a brake. Braking occurs in quadrants II and IV. (See Figure 13–15.)

Optional Braking Resistor

In order for an AC drive to operate in quadrant II or IV, a means must exist to deal with the electrical energy returned to the drive by the motor. Electrical energy returned by the motor can cause voltage in the DC link to become excessively high when added to existing supply voltage. Various drive components can be damaged by this excessive voltage. The braking resistor is connected to terminals B+ and B–. The braking resistor is added and removed from the circuit by an IGBT. Energy returned by the motor is seen on the DC link. When the DC link reaches a predetermined limit, the IGBT is switched on by the control logic. The resistor is placed across the DC link. Excess energy is dissipated by the resistor, reducing bus voltage. When DC link voltage is reduced to a safe level the IGBT is switched off, removing the resistor from the DC link. This is referred to as *pulsed resistor braking*. This process allows the motor to act as a brake, slowing the connected load quickly (Figure 13–16).

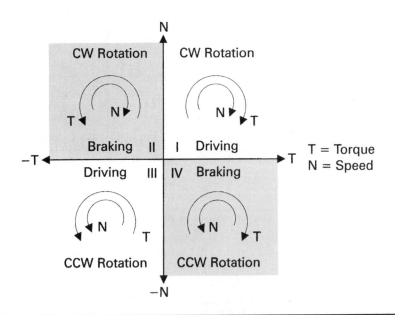

Figure 13–15 Four-quadrant operation. Courtesy Siemens Energy and Automation, Inc.

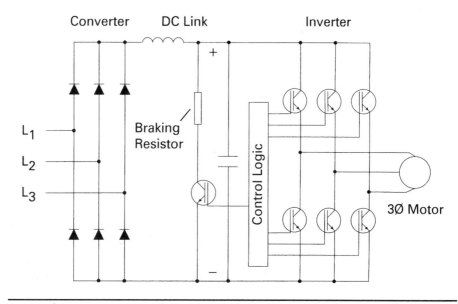

Figure 13–16 Optional braking with resistor. Courtesy Siemens Energy and Automation, Inc.

When applying an AC drive and motor to an application it is necessary to know the horsepower, torque, and speed characteristics of the load. The following categorizes and shows characteristics of various loads.

Constant torque	The load is essentially the same throughout the speed range. Examples are hoisting gear and belt conveyors.
Variable torque	The load increases as speed increases. Examples are pumps and fans.
Constant horsepower	The load decreases as speed increases. Examples are winders and rotary cutting machines.

Constant Torque Applications

A constant torque load implies that the torque required to keep the load running is the same throughout the speed range. It must be remembered that constant torque refers to the motor's ability to maintain constant flux (Φ). Torque produced will vary with the required load. One example of a constant torque load is a conveyor similar to the one shown in Figure 13–17. Conveyors can be found in all sorts of applications and environments, and can take many styles and shapes.

Conveyors are made up of belts to support the load, various pulleys to support the belt, maintain tension, and change belt direction, and idlers to support the belt and load.

Determining Motor Speed

The speed and horsepower of an application must be known when selecting a motor and drive. Given the velocity in feet per minute (FPM) of the conveyor belt, the diameter in inches of the driven pulley, and the gear ratio (G) between the motor and driven pulley, the speed of the motor can be determined. The following formula is used to calculate conveyor speed.

$$Motor\ RPM = \frac{Conveyor\ Velocity\ (FPM) \times G}{\pi \times (Diameter\ in\ Inches \div 12)}$$

If, for example, the maximum desired speed of a conveyor is 750 FPM, the driven pulley is 18 inches in diameter, and the gear ratio between the motor and driven pulley is 4:1, the maximum speed of the motor is 638.3 RPM. It would be difficult to find a motor that would operate at exactly this speed. An AC drive can be used with an eight-pole motor (900 RPM). This would allow the conveyor to be operated at any speed between zero and the desired maximum speed of 750 FPM.

$$Motor\ RPM = \frac{750 \times 4}{3.14 \times (18 \div 12\ or\ 1.5)}$$

$$Motor\ RPM = 637\ RPM$$

Another advantage to using AC drives on a conveyor is the ability to run different sections of the conveyor at different speeds. A bottle machine, for example, may have bottles bunched close together for filling and then spread out for labeling. Two motors and two drives would be required. One motor would run the filling section at a given speed and a second motor would run the labeling section slightly faster, spreading the bottles out. (See Figure 13–17.)

Horsepower

Calculating motor horsepower is complicated with many variables, which is beyond the scope of this course. Someone with knowledge of and experience with conveyor operation would be required to accurately calculate the required horsepower. The horsepower required to drive a conveyor is the effective tension (Te) times the velocity (V) of the belt in feet per minute, divided by 33,000.

$$HP = \frac{Te \times V}{33,000}$$

Effective tension (Te) is determined by several forces:

- Gravitational weight of the load
- Length and weight of belt
- Friction of material on the conveyor

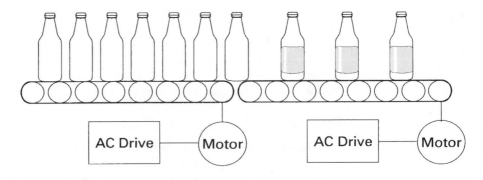

Figure 13–17 Pictorial of AC drive operation. Courtesy Siemens Energy and Automation, Inc.

- Friction of all drive components and accessories
  Pulley inertia
  Belt/chain weight
  Motor inertia
  Friction of plows
  Friction of idlers
- Acceleration force when new material is added to conveyor

If the effective tension of a conveyor were calculated to be 2000 pounds and the maximum speed is 750 FPM, then the required horsepower is 45.5, as illustrated by the following:

$$HP = \frac{2000 \times 750}{33,000}$$
$$HP = 45.5$$

Starting torque of a conveyor can be 1.5 to 2 times full load torque. A motor capable of driving a fully loaded conveyor may not be able to start and accelerate the conveyor up to speed. Alternating current drives can typically supply 1.5 times full load torque for starting. In order to start and accelerate the conveyor, an engineer may need to choose a larger motor and drive.

Torque/HP and Speed

The speed on a conveyor is increased by increasing the AC drive frequency (F) to the motor. Torque (T) is affected by flux (Φ) and working current (Iw). The drive will maintain constant flux by keeping the voltage and frequency ratio constant. To do this the drive increases voltage and frequency in proportion. During acceleration, working current increases, causing a corresponding increase in torque. Once it reaches its new speed, the working current and torque will be the same as at its old speed. The conveyor cannot be operated above the rated frequency of the motor (60 Hz) because the drive will no longer be able to provide constant flux. As a result, the motor will be unable to supply rated torque. Horsepower (HP) is affected by torque and speed. There will be a corresponding increase in horsepower as speed (RPM) increases.

Variable Torque Applications

A variable torque load implies that torque and horsepower increase with an increase in speed. Fans and pumps are examples of variable torque.

Variable Torque Pumps

There are several types of pumps. The most common pump is the end-suction centrifugal pump. There are variations of the centrifugal pump. Turbine and propeller pumps are examples. This section deals with variable torque loads. The faster a centrifugal pump turns, the more fluid it pumps and the more torque it requires. It should be noted that not all pumps are variable torque. Reciprocating pumps are constant torque.

Calculating Horsepower

Calculating horsepower for a pump application is an involved process that requires someone with a thorough knowledge of the application and pumps. The following information is for illustration only. There are three related horsepower calculations involved in pump applications: liquid, mechanical, and electrical.

Liquid Horsepower *Liquid horsepower* is the hydraulic power transferred to the pumped liquid. The following formula can be used to calculate liquid energy.

Liquid Energy (in ft-lb) = Total Head × (Gallons × Weight)

Water weighs 8.34 pounds per gallon. If 50 gallons of water per minute were required to be moved through 100 feet of head the energy required would be 41,700 ft-lb/minute.

100 feet × (50 gallons × 8.34) = 41,700 ft-lb/minute

If the pump's speed was increased so that 100 gallons of water was being pumped through 100 feet of head, the energy would be 83,400 ft-lb/minute. Twice the energy would be required. The hydraulic head would, in actuality, also increase.

100 feet × (100 gallons × 8.34) = 83,400 ft-lb/minute

The common method of expression is horsepower (HP). One horsepower is equal to 33,000 ft-lb/minute. Therefore, 41,700 ft-lb/minute is 1.26 HP and 83,400 ft-lb/ minute is 2.53 HP, as illustrated by:

$$\frac{41,700}{33,000} = 1.26 \; HP \qquad \frac{83,400}{33,000} = 2.53 \; HP$$

Mechanical Horsepower *Mechanical horsepower* is the horsepower input to the pump and is equal to the liquid horsepower divided by the pump's efficiency. If the liquid horsepower is 2.53 and the pump is 75 percent efficient, the brake horsepower is 3.4 HP.

$$\frac{1.26}{.75} = 1.75 \; HP \qquad \frac{2.53}{.75} = 3.4 \; HP$$

Electrical Horsepower *Electrical horsepower* is the horsepower required to run the motor driving the pump and is equal to the mechanical horsepower divided by the motor's efficiency. If the motor is 90 percent efficient the electrical horsepower is 3.78 HP. It can be seen that with an increase of pump speed there is a corresponding increase in electrical horsepower.

$$\frac{1.7}{1.9} = .89 \; HP \qquad \frac{3.4}{.90} = 3.8 \; HP$$

Hydraulic Head

Hydraulic head is the difference in hydraulic pressure between two points, which actually includes elevation, pressure, and velocity. An increase in pump speed would cause increases in pressure and velocity, and in turn increases the hydraulic head.

Torque/HP and Speed

The speed on a pump is increased by increasing the AC drive frequency (F) to the motor. Torque (T) is affected by flux (Φ) and working current (Iw). The drive will maintain appropriate flux by adjusting the voltage and frequency ratio dependent on speed. During acceleration, working current will increase, causing a corresponding increase in torque. In this application, torque increases in proportion to the speed squared. This is due to the increase in hydraulic head as the pump works harder to pump more fluid. Horsepower increases in proportion to the speed cubed due to an increase of torque and speed. The pump cannot be operated above the rated frequency of the motor (60 Hz) because the drive will no longer be able to provide constant flux. As a result, the motor will be unable to supply rated torque. The load's torque requirements increase while the motor's ability to supply torque decreases.

Fans

This same principal applies to fan applications. The horsepower of a fan is determined by dividing the product of air flow (in cubic feet per minute) and pressure by the product of the constant 6356 and fan efficiency. Increasing the speed of the fan increases air flow and pressure, requiring the motor to work harder (Iw increases). Torque and horsepower increase.

$$HP = \frac{Flow \times Pressure}{6356 \times fan\ efficiency}$$

Constant Horsepower Applications

Constant horsepower applications require a constant force as radius changes. A lathe, for example, starts out with a certain diameter object. As the object is cut and shaped the diameter is reduced. The cutting force must remain constant. Another example of a constant horsepower application is a winder where radius increases as material is added to a roll and decreases as material is removed.

Relationship of Speed, Torque, and Horsepower

Applications such as lathes, that are driven in a continuous circular motion, are sometimes referred to as spindle drives. Horsepower will remain constant in a spindle drive application. The surface speed in feet per minute (FPM) is equal to *2n* times the radius (in feet) of the material times the speed in RPM. Surface speed will remain constant as the material is shaped and the radius reduced. Torque is equal to force times radius. Horsepower is equal to torque times speed.

$$Surface\ Speed\ (FPM) = 2n \times Radius \times Speed\ (RPM)$$
$$Torque = Force \times Radius$$
$$HP = Torque \times Speed$$

Troubleshooting Inverter Drives

The Four Main Areas of Possible Problems

In an AC inverter drive the four main areas of possible problems are the same as those in a DC drive:

1. The electrical supply to the motor and the drive.
2. The motor and/or its load.
3. The feedback device and/or sensors that provide signals to the drive.
4. The drive itself.

Even though the problem may be in any one or several of these areas, the best place to begin troubleshooting is within the drive unit itself. The reason for this is that most drives have some type of display that aids in troubleshooting. This display may be simply an LED that illuminates to indicate a specific fault condition, or it may be an error or fault code that can be looked up in the operator's manual. For this reason, it is strongly suggested that a copy of the fault codes be made and fastened to the inside of the drive cabinet, where it will be readily available to the maintenance technician. The original should be placed in a safe location such as the maintenance supervisor's office.

Before you proceed, *stop and think about what you are doing!* Before working on any electrical circuit, *remove all power.* Sometimes removing the power is not possible or permissible. In these instances, *work carefully and wear the appropriate safety equipment.* Do not rely on safety interlocks, fuses, or circuit breakers to provide personal protection. *Always verify that the equipment is deenergized, and tag and lock the circuit out!*

Even when the power has been removed, you are still subject to shock and burn hazards. Most drives have high-power resistors inside, and these resistors can and do get hot! Give them time to cool down before touching them. Most drives also have large electrolytic capacitors. These capacitors can and do store an electrical charge. Usually the capacitors have a bleeder circuit that dissipates this charge. However, this circuit may have failed. *Always verify that electrolytic capacitors are fully discharged.* Do this by measuring carefully any voltage present across the capacitor terminals. If voltage is present, *use an approved shorting device to discharge the capacitor completely* .

Now slow down, take your time, and use your senses, even though production has stopped, and you are under pressure to fix the equipment. Sometimes a little extra initial time to take stock of the situation can save considerable time later. So do the following:

Look! Do you see any charred or blackened components? Have you noticed any arcing? Do fuses or circuit breakers appear blown or tripped? Do you see any discoloration around wires, terminals, or components? A good visual inspection can save a lot of troubleshooting time.

Listen! Did you hear any funny or unusual noises? A frying or buzzing sound may indicate arcing. A hum may be normal or an indication of loose laminations in a transformer core. A rubbing or chafing sound may indicate that a cooling fan is not rotating freely.

Smell! Do you notice any unusual odors? Burnt components and wires give off a distinctive odor; metal will smell hot if subjected to too much friction.

Touch! (Be very careful with this!) If components feel cool, that may indicate that no current is flowing through the device. If components feel warm, chances are that everything is normal. If components feel hot, it could be an indicator that either too much current is flowing, or too little cooling is taking place. In any event, there may be a problem worth further investigation.

The point of all this is that by being *observant* you have a good chance of discovering the problem or problem area. Before we consider in more detail the four main areas mentioned earlier, review the proper techniques for installing an inverter.

Now let's examine some of the parameters that can be programmed into an inverter. Keep in mind that not all inverters share these parameters. Some manufacturers list the same parameters in different order. Certain manufacturers have other names for the settings discussed here. Also, keep in mind that some inverters will have more parameters and others will have fewer parameters than those mentioned. Again, familiarize yourself with the particular inverter that you are using. In general, programmable inverter parameters include the following.

Frequency   The output frequency of the inverter.

Voltage   The output voltage of the inverter.

Current   The output current of the inverter.

Torque   Calculated motor torque. Dependent on the programmed settings of the motor, nominal current, and the motor magnetization.

Basic setup   Can be set for constant torque or constant speed.

Local/remote   Can be set for local control from the inverter's control panel or remote control from a start/stop station *located away from the inverter* cabinet.

Minimum speed   Slowest speed at which the motor will run.

Maximum speed   Depending on the setting, it may be possible to attain a speed higher than the rated speed of the motor. This parameter must be set higher than the minimum speed setting. If it is set lower than the minimum speed, the motor will not run.

Jogging speed   Sets the speed of the motor when jogging.

Ramp time up   This is the acceleration time (the time required to get from minimum speed to maximum speed) expressed in seconds. If this time is too short, the inverter can trip out.

Ramp time down   This is the deceleration time (the time required to get from maximum speed to minimum speed) expressed in seconds. If this time is set too short, the inverter can trip out.

Current limit   Sets the maximum current available to the motor. If the setting is at the maximum permissible value, the motor will have maximum starting torque. This value can be on the order of 160 percent of nominal motor current. If the current limit is set too low, the inverter can trip out.

Motor power   This is the motor's power rating expressed either in horsepower (HP) or in kilowatts (kw). Some drives will accept the value in either unit of measurement while others require converting the units from one measure to the other.

Motor nominal voltage   The rated line voltage of the motor from the motor nameplate. This value should be set as closely as possible to the specified value.

Motor nominal frequency   The rated frequency of the motor from the motor nameplate. This value should be set as closely as possible to the specified value.

Start voltage   Typically, the factory setting for this value should be adequate. This setting is affected by the motor power, motor nominal voltage, and motor nominal frequency values.

Start compensation   Typically, the factory setting for this value should be adequate. This setting is affected by the motor power, motor nominal voltage, and motor nominal frequency values.

Slip compensation   Typically, the factory setting for this value should be adequate. This setting is affected by the motor power, motor nominal voltage, and motor nominal frequency values.

V/f ratio   Typically, the factory setting for this value should be adequate. This setting is affected by the motor power, motor nominal voltage, and motor nominal frequency values.

Start/stop mode   This value is programmed for the various types of start/stop circuits used. For example, a two-wire start/stop, a three-wire start/stop, a three-wire start/stop with a jog, and so on.

Digital input select   If selected, this value bypasses the ramp time down setting, and the inverter decelerates the motor in the shortest possible amount of time as a result.

Analog input select   Sets either a 0-10V or 4-20 mA reference that is proportional to the frequency input.

Relay output select   Provides a contact closure when the inverter is placed in the run mode.

Analog output select   Provides a 4-20 mA output signal that is proportional to the motor speed.

Thermal motor protect   Depending on the setting chosen, this parameter will either flash the display when the motor's critical temperature is reached or trip the inverter.

Trip reset mode   If selected, this parameter will prevent the inverter from restarting automatically after a trip.

DC brake time   If selected, this parameter will provide additional braking torque at low motor speeds.

Motor nominal current   Set to the full load current rating on the motor's nameplate.

Motor magnetization   Set to the no load current rating on the motor's nameplate.

Where to Begin Troubleshooting an Inverter Drive System

In order to begin troubleshooting, we need to understand the general, step-by-step sequence of events that most inverter drive systems follow on startup and shutdown.

We begin with the AC power applied to the inverter. We must adjust the reference or set point for the desired speed and torque characteristics of the motor. Next, the start/stop circuit is placed in the start or run mode. Instantly, the main control components begin a diagnostic routine. We will examine some of the routine's fault codes shortly. If no faults are detected, the driver activates the power semiconductors. These produce the output frequency and V/f ratio programmed into the inverter to match the speed and torque settings. The programmed setting for the ramp time up will control how long it takes the motor to reach the desired speed. While the motor speed is ramping up, the motor current is monitored. Should the current exceed the programmed current limit setting, the inverter may (depending on the manufacturer) automatically adjust the ramp time up program or simply trip. If tripping occurs repeatedly, the ramp time up program may need to be modified to accommodate a longer acceleration time. The ramp time up setting will cause the inverter output voltage and frequency to increase, accelerating the motor to the desired speed set point.

When it becomes necessary to stop the motor, most inverters offer several options. One option is basically to let the motor coast to a stop. This can take a very long time for high-inertia loads. The ramp down time setting allows the inverter to slow the motor gradually. This deceleration is accomplished by allowing the motor to feed its self-generated energy back into the inverter. The inverter will use large resistors to absorb this energy. This process, called dynamic braking, is usually insufficient to bring the motor to a controlled and rapid stop because as the motor slows, less energy is self-generated. Therefore it is common practice to use a mechanical brake in conjunction with dynamic braking to provide additional braking action at slow speeds. Another method of stopping the motor without using a mechanical brake is plugging or DC injection. When the operator wishes to stop the motor, DC current is fed into the motor winding. This current replaces the rotating magnetic field with a fixed magnetic field. The rotor soon becomes locked with this fixed magnetic field, which in effect stops the motor rotation. This method should not be used repeatedly because heat can build up in the motor as a result and damage the windings. Now that we have a basic understanding of the processes that an inverter follows on startup and shutdown, let's investigate the types of problems that can be encountered in the four main areas of an inverter drive system.

The Electrical Supply to the Motor and the Drive

Most maintenance technicians believe that the power distribution in an industrial environment is reliable, stable, and free of interference. Nothing could be further from reality! Frequent outages, voltage spikes and sags, and electrical noise are normal operating occurrences. The effect of these is not as detrimental to motor performance as it can be to the operation of the drive itself. Most AC inverters are designed to operate despite variations in supply voltages. Typically, the incoming power can vary as much as ±10 percent with no noticeable change in drive performance. However, in the real world it is not unusual for power line fluctuations to exceed 10 percent. These fluctuations may occasionally cause a controller to trip. If tripping occurs repeatedly, a power line regulator may be required to hold the power at a constant level.

Should the power supply to the controller fail, a power line regulator will be of little use. In this situation, an uninterruptible power supply (UPS) is needed. Several manufacturers produce a complete power line conditioning unit. These units combine a UPS with a power line regulator.

Quite often controllers are connected to an inappropriate supply voltage. For example, it is not unusual for a drive rated 208V to be connected to a 240V supply. Likewise, a 440V rated drive may be connected to a 460V or even a 480V source. Usually the source voltage should not exceed the voltage rating of the drive by more than 10 percent. For a drive rated at 208V the maximum supply voltage is 229V (208 × 10% (20.8) + 208 = 229). Obviously the 208V drive, when connected to the 240V supply, is receiving excess voltage and should not be used. For our 440V drive the maximum supply voltage is 484V (440 × 10% (44) + 440 = 484). Although this value appears to fall within permissible limits, another potential problem exists here. Suppose that the power line voltage fluctuates by 10 percent. If the 440V source suffers a 10 percent spike, the voltage will increase to 484V. This value is within the design limits of the drive. But what happens when we connect the 440V drive to a 460V or a 480V power line? If we experience that same 10 percent spike, the 460V line will increase to 506V (460 × 10% (46) + 460 = 506), and the 480V line will increase to 528V (480 × 10% (48) + 480 = 528). We have thus exceeded the voltage rating of the drive and probably damaged some internal components! Most susceptible to excess voltage and spikes or transients are the SCRs, MOSFETs, and power transistors. Premature failure of capacitors can also occur. As you can see, it is very important to match the line voltage to the voltage rating of the drive.

An equally serious problem occurs when the phase voltages are unbalanced. Typically, during construction care is taken to balance the electrical loads on the individual phases. As time goes by and new construction and remodeling occurs, it is not unusual for the loading to become imbalanced, causing intermittent tripping of the controller and perhaps premature failure of components. To determine if an imbalanced phase condition exists you will need to do the following:

1. Measure and record the phase voltages (L_1 to L_2, L_2 to L_3, and L_1 to L_3).
2. Add the three voltage measurements from Step 1 and record the sum of all phase voltages.
3. Divide the sum from Step 2 by 3 and record the average phase voltage.
4. Now subtract the average phase voltage obtained in Step 3 from each phase voltage measurement in Step 1 and record the results. (Treat any negative answers as positive answers.) These values are the individual phase imbalances.
5. Add the individual phase imbalances from Step 4 and record the total phase imbalance.
6. Divide the total phase imbalance from Step 5 by 2 and record the adjusted total phase imbalance.
7. Now divide the adjusted total phase imbalance from Step 6 by the average phase voltage from Step 3 and record the calculated phase imbalance.
8. Finally, multiply the calculated phase imbalance from Step 7 by 100 and record the percent of total phase imbalance.

Consider this example involving a 440V three-phase supply to an AC inverter drive to see how this process works.

1. L_1 to L_2 = 432 V; L_2 to L_3 = 435 V; and L_1 to L_3 = 440 V.
2. The sum of all phase voltages equals 432 V + 435 V + 440 V, or 1307 V.
3. The average phase voltage is equal to 1307 V divided by 3, or 435.7 V.
4. To find the individual phase imbalances, we subtract the average phase voltage from the individual phase voltages and treat any negative values as positive. So L_1 to L_2 = 432 V minus 435.7 V, or 3.7 V; L_2 to W = 435 V, 435.7 V, or 0.7 V; and L_1 to L_3 = 440 V minus 435.7 V, or 4.3 V.
5. Now we find the total phase imbalances by adding the individual phase imbalances:

$$3.7 \text{ V} + 0.7 \text{ V} + 4.3 \text{ V} = 8.7 \text{ V}$$

6. To find the adjusted total phase imbalance we divide the total phase imbalance by 2:

$$\frac{8.7 \text{ V}}{2} = 4.35 \text{ V}$$

7. Next we find the calculated phase imbalance by dividing the adjusted total phase imbalance by the average phase voltage:

$$\frac{4.35 \text{ V}}{435.7 \text{ V}} = 0.00998$$

8. Finally we multiply the calculated phase imbalance by 100 to find the percent total phase imbalance:

$$0.00998 \times 100 = 0.998\%$$

In this example we are within tolerances, and the differences in the phase voltages should not cause any problems. In fact, as long as the percent total phase imbalance does not exceed 2 percent, we should not experience any difficulties as a result of the differences in phase voltages.

What Problems Can Occur with the Motor and the Load?

Probably the most common cause of motor failure is heat. Heat can occur simply as a result of the operating environment of a motor. Many motors are operated in areas of high ambient temperature. If steps are not taken to keep the motor within its operating temperature limits, the motor will fail. Some motors have an internal fan that cools the motor. If the motor is operated at reduced speed, this internal fan may not turn fast enough to cool the motor sufficiently. In these instances, an additional external fan may be needed to provide additional cooling to the motor. Typically, such fans are interlocked with the motor operation in such a way that the motor will not operate unless the fan operates as well. Therefore a fault in the external fan control may prevent the motor from operating.

The sensors used to sense motor temperatures consist simply of a nonadjustable thermostatic switch that is normally closed and opens when the temperature rises to a certain level. In this event you must wait for the motor to cool down sufficiently before resetting the temperature sensor and restarting the motor.

Periodic inspection of the motor and any external cooling fans is strongly recommended. The fans should be checked for missing or bent vanes. All openings for cooling should be kept free of obstructions. Any accumulation of dirt, grease, or oil should be removed. If filters are used, these must be cleaned or replaced on a routine basis.

Heat may also cause other problems. When motor windings become overheated, the insulation on the wires may break down. This breakdown may cause a short, which may lead to an open condition. A common practice used to find shorts or opens in motor windings is to megger the windings with megohm meter. Extreme caution must be taken when using a megger on the motor leads. Be certain that you have disconnected the motor leads from the drive. Failure to do this will cause the megger to apply a high voltage to the output section of the drive. Damage to the power semiconductors will result. You may also decide to megger the motor leads at the drive cabinet. Again, be certain that you have disconnected the leads from the drive unit and megger only the motor leads and motor winding. *Never megger the output of the drive itself!*

Since the motor drives a load of some kind, it is also possible for the load to create problems. The drive may trip out if the load causes the motor to draw an excessive amount of current for too long a time. When this occurs, most drives display some type of fault indication. The problem may be a result of the motor operating at too high a speed. Quite often, a minor reduction in speed is all that is necessary to prevent repeated tripping of the drive. The same effect occurs if the motor is truly overloaded. Obviously, in this case either the motor size needs to be increased or the size of the load decreased to prevent the drive from tripping.

Some loads have a high inertia. They require not only a large amount of energy to move, but once moving, a large amount of energy to stop. If the drive cannot provide sufficient braking action to match the inertia of the load, the drive may trip, or overhaul. A drive with greater braking capacity is needed in such cases to prevent tripping from recurring.

Can Feedback Devices or Sensors Cause Problems?

Mechanical vibration may cause the mounting of feedback devices to loosen and their alignment to vary. Periodic inspections are necessary to verify that these devices are aligned and mounted properly.

It is also important to verify that the wiring to these devices is in good condition and the terminations are clean and tight. Another consideration regarding the wiring of feedback devices is electrical interference. Feedback devices produce low-voltage/low-current signals that are applied to the drive. If the signal wires from these devices are routed next to high-power cables, interference can occur. This interference may result in improper drive operation. To eliminate the possibility that this will occur, several steps must be taken. The signal wires from the feedback device should be installed in their own conduits. Do not install power wiring and signal wiring in the same conduit. The signal wires should be shielded cable, with the shield wire grounded to a good ground at the drive cabinet only. Do not ground both ends of a shielded cable. When the shielded signal cable is routed to its terminals in the drive cabinet, the cable should not be run or bundled parallel to any power cables, but instead at right angles to such power cables. Furthermore, the signal cable should not be routed near any highpower contactors or relays. When the coils of a contactor or relay are energized and deenergized, a spike is produced. This spike can create interference with the drive. To suppress this spike, it may be necessary to install a freewheeling diode across any DC coils or a snubber circuit across any AC coils.

Problems That Can Occur in the Inverter Drive

First, look for fault codes or fault indicators. Most drives provide some form of diagnostics, and this can be a great timesaver.

Let's look at some of the fault codes, symptoms, probable causes, and fixes for common problems:

- *The inverter is inoperable, and no LED indicators are illuminated.* It's possible that no incoming power is present. You can verify this by measuring the voltage at the power supply input terminals in the inverter cabinet. The problem may be caused by a blown fuse, an open switch, an open circuit breaker, or an open disconnect. There are several things to check if the fuses are blown. One item to look for is a shorted metal oxide varistor (MOV), a device that provides surge protection to the inverter. If a significant power surge occurred, the MOV may have shorted to protect the inverter. Because the MOV is located across the power supply lines (to provide protection), a shorted MOV can cause fuses to blow. Another reason why fuses blow is a shorted diode in the rectifier circuit. A shorted or leaky filter capacitor in the power supply may also cause fuses to blow.

- *The inverter is powered up but does not work. There are indications of a fault condition.* The "watchdog" circuit may have tripped. Remember that a watchdog circuit monitors the power lines for disturbances, and if the disturbance occurs for a long duration, the watchdog circuit trips the controller to protect it from damage. A heavy starting load may sag the power line voltage to such a point that the inverter receives insufficient voltage. This deficiency can trip the inverter. Likewise, if the load has high inertia, it is possible for the regenerative effect to provide excess voltage to the inverter. Another way that the inverter may receive excess voltage is use of the power factor correction capacitors while the load is removed.

If the watchdog circuit has not tripped, other possible reasons exist for this fault condition. It is possible that there are interlocks on the cabinet or cooling fans, and one or more of these may be open. Likewise, a temperature sensor on a heatsink on the power semiconductors in the inverter or in the motor itself may have detected an excessive temperature condition and opened as a result.

- *The inverter is energized, and a fault is indicated. The motor does not respond to any control signals.* If the load is too high for the motor settings, the motor may fail to rotate. It may be necessary to increase the current limit or voltage boost settings to allow the motor to overcome the load. Another possibility is that the load is overhauling the motor. If this is the situation, it will be necessary to adjust the deceleration time to allow the motor to take longer to brake the load. It may also be necessary to add auxiliary braking in the form of a mechanical brake. If the motor leads have developed a short, or the motor itself has a shorted winding or is overloaded or stalled, the current limit sensor may trip the inverter. Tripping may also occur as a result of a shorted power semiconductor.

As you can see, you need to be aware of many areas when dealing with an inoperable inverter. Fortunately, the inverter itself can help a great deal by displaying fault codes. The operator's manual will interpret the fault codes and give instructions for clearing the fault condition. Let's examine some fault codes and how they can help in your troubleshooting. Remember that not all manufacturers provide the same fault codes. Some provide more, and others provide fewer.

As mentioned earlier, one fault that may be displayed indicates an overcurrent limit condition. This indicator should direct you to examine the motor for mechanical binding, jams, and so forth. To verify whether one of these conditions is the cause of the problem, disconnect the motor from the load and reset the inverter. If the fault clears, then you know that the load is the cause. If the fault reappears, you need to look further, perhaps at the motor itself.

Another fault code, over voltage, may be the result of a high-inertia load that causes overhauling. This fault code may also be a result of setting the deceleration ramp down parameter for rapid deceleration. Lengthening the deceleration time may clear the fault. If lengthening the time is not possible, additional mechanical braking may be required to bring the load to a rapid stop.

The inverter overload fault code is an indication of electrical problems. Examples of these are shorted or grounded motor leads or windings and/or defective power semiconductors. If the motor is suspect, disconnect it from the inverter. If the fault clears when you reset the inverter, you can assume that the problem lies in the motor and/or its leads. To verify whether the problem is in the power semiconductors, disconnect the gate lead from one of the devices. Reset the inverter. If the fault clears, you have found the problem. If the fault is still present, reconnect the gate lead, move to the next device, and disconnect its gate lead. Reset the inverter. Again, if the fault clears, you have found the problem. If the fault is still present, repeat the preceding steps until you have tested all of the power semiconductors.

Another fault code indicates shorted control wiring. If this code is displayed, simply disconnect the control wiring and reset the inverter. The fault should clear. This result indicates problems in the control wiring. If the fault does not clear, try unplugging the control board and resetting the inverter. A cleared fault condition in this case indicates problems in the control board.

It is a good idea to maintain an inventory of spare PC boards for the various inverters at your plant. That way, if you determine that the problem is caused by a defective PC board, it should be fairly simple to replace the board with a spare. This will minimize downtime and allow you to try to repair the PC board in the shop, under less pressure. If the PC board cannot be repaired, it may be possible to return it to the manufacturer for repair or exchange.

Although fault codes are excellent troubleshooting tools and can save a great deal of time, you should be aware of other potential problems that may or may not show up as fault codes, depending on the manufacturer.

Heat can produce problems in the drive unit. The cabinet may have one or more cooling fans, with or without filters. These fans are often interlocked with the drive power in such a way that the fan must operate in order for the drive to operate. Make certain that the fans are operational and the filters are cleaned or replaced regularly. The power semiconductors are typically mounted on heatsinks. A heatsink may have a small thermostat mounted on it to detect an excess temperature condition in the power semiconductors. If the heatsink becomes too hot, the thermostat will open, and the drive will trip. Usually these thermostats are self-resetting. You must wait for them to cool down and reset themselves before the drive will operate. If tripping occurs repeatedly, a more serious problem exists that requires further investigation.

If the drive is newly installed, problems are often the result of improper adjustments to the drive. On the other hand, if the drive has been in operation for some time, it is unlikely that readjustments are needed. All too often, an untrained individual will try to adjust a setting to "see if this fixes it." Usually such adjustments only make things worse. This is not to say that adjustments are never needed. For example, if the process being controlled is changed or some component of the equipment has been replaced, it probably will be necessary to change the drive settings. For this reason it is very important to record the initial settings and any changes made over the years. This record should be placed in a safe location and a copy made and placed in the drive cabinet for easy access by maintenance personnel.

Replacing the drive should be the last resort. More often than not, if the drive is defective, something external to the drive is the reason for the drive's failure. Replacing the drive without determining what caused the failure may result in damage to the replacement drive. If the drive has failed, the possibility still exists that you can get it to work again.

Most drive failures occur in the power section, where you will find the SCRs, MOSFETs, and so on. These devices can be tested with reasonable accuracy with nothing more than an ohmmeter. (See Appendix F.) If you determine that one or more of these devices is defective, you can usually obtain a substitute part from a local electronics parts supplier. If a part is not available, the drive will have to be returned to the manufacturer for service or an on-site visit from a field service technician will need to be scheduled.

If it is determined that the problem is not in the power section, then it must be located in the control section of the drive. The electronics used in the control section are more complex, and therefore troubleshooting is not recommended. In this event, the drive must be returned to the manufacturer for repair or arrangements made for on-site repair by a factory-trained technician.

CHAPTER 14

Motor Control Center

Storage

If the motor control center cannot be placed into service soon after its receipt, it should be stored in a clean, dry, and ventilated building free from temperature extremes. Acceptable storage temperatures are from 0° C (32° F) to 40° C (104° F).

If the storage area is cool and/or damp, enough heat should be provided to prevent condensation, which could harm the motor control center.

NOTE: Outdoor storage is inadequate, even with the protection of a tarpaulin.

Handling

Adequate equipment for handling the motor control center, such as a fork lift truck, crane, or rods and pipe rollers, is necessary. The following chart, with approximate weights of single sections full of typical units, is a guide for determining the type of handling equipment needed.

Enclosure Types	Weight (in pounds) by Number of Sections		
	1	2	3
NEMA 1, 1A, 12–15″ Deep	600	1200	1800
NEMA 3R Non Walk-In, 20.5″ Deep	900	1800	2700
NEMA 1, 1A, 12–20″ Deep	750	1500	2250
NEMA 3R, Non Walk-In, 25.5″ Deep	1050	2100	3150

Weights vary by enclosure type and depth. A single skid could contain from one to three vertical sections. The following instructions are provided to avoid injury and equipment damage while moving the motor control centers:

1. Caution should be exercised when moving heavy equipment.

2. Verify the capabilities of the moving equipment to handle the weight.

3. Fork trucks, when available, provide a quick and convenient method of moving motor control centers.

4. Lifting angles are provided on each shipping block for handling the motor control center by overhead cranes. The following precautions should be taken when using a crane:
   (a) Handle in the upright position only.
   (b) Select rigging lengths to compensate for any unequal weight distribution.
   (c) Do not exceed the 45° maximum between the vertical and lifting cables.
   (d) Use only slings with safety hooks or shackles. Do not pass ropes or cables through the holes in the lifting angle.

5. Use extreme caution when moving sections with rods or pipe rollers. There is a tendency for the control center to tilt due to its high center of gravity.

After the shipping section is in place, its lifting angle may be removed and discarded. To prevent entrance of foreign matter, replace all hardware that was used to secure the lifting angle.

NOTE: Do not attempt to attach lifting means to sections provided with pull boxes or to lift sections provided with pull boxes.

Installation

Motor control centers are not designed to be placed in hazardous locations. The area chosen should be well-ventilated and free from excess humidity, dust, and dirt. The temperature of the area should be no less than 0° C (32° F) and no greater than 40° C (104° F). For indoor locations, protection from moisture or water entering the enclosure must be provided.

Motor control centers should be located in an area which allows a minimum of three feet of free space in front of front-of-board construction. An additional three feet should be allowed in the rear of back-to-back construction. This free space will provide adequate room for removing and installing units. A minimum 1/2 inch space should be provided between the back of front-of-board motor control centers and a wall (6 inches for damp locations).

When selecting a location for the installation of a motor control center, careful attention should be given to accessibility, overhead clearances, and future expansions. Consideration of these factors will eliminate many difficulties during the installation of this and future motor control centers.

Motor control centers are assembled in a factory on smooth, level surfaces to ensure that all sections are properly aligned. A similar smooth and level surface should be provided by the customer for installation. An uneven foundation may cause misalignment of shipping blocks, units, and doors. The surface under a motor control center must be of a noncombustible material unless bottom plates are installed in each vertical section.

Joining Sections

WARNING: There is a hazard of electrical shock or burn whenever working in or around electrical equipment. Turn off the power supply to the equipment before working on it.

Before positioning the motor control center sections, check for damaged bus bars and insulators. If the bus is bent or the insulators are broken, do not install the motor control center. Report any damage to the carrier. See the manufacturer's specifications for details before installing new sections.

Ground Bus Splicing: See Manufacturer's Specifications
Conductor Entry: See Manufacturer's Specifications
Load and Control Wiring: See Manufacturer's Specifications

Pre-Operational Check List

To ensure proper operation of the motor control center the items listed below should be checked *before* energizing the motor control center:

1. Complete the maintenance procedure described later in this text. This initial maintenance is extremely important for detecting shipping damage or loose connections. The motor control center should not be energized until initial maintenance is complete.

2. Perform insulation test on the motor control center (see maintenance procedure described later in this text).

3. If the motor control center is equipped with ground fault protection, this should be adjusted properly before energizing.

4. All blocks or other temporary holding means should be removed from the electrical devices.

5. Current transformers should have the secondary shunt bar removed. Do not operate a current transformer with its secondary open circuited.

6. Manually exercise all switches, circuit breakers, and other operating mechanisms to make sure they are properly aligned and operating freely.

7. Electrically exercise all electrically-operated switches, circuit breakers, and other mechanisms (but not under load) to determine that the device operates properly. An auxiliary source of control power may be required.

8. Timers should be checked for the proper interval and contact operation.

9. Check overload selection tables against motor full load current to ensure that proper overload units have been installed.

10. Make certain all load and remote control connections have been made and agree with wiring diagrams provided.

11. Make certain that all ground connections are made. Make certain all load and remote control connections fit properly.

12. Install covers and close doors and make certain they are properly tightened.

Energizing Equipment

WARNING: Energizing a motor control center for the first time is potentially dangerous. Therefore, only qualified personnel should energize the equipment. If faults caused by damage or poor installation practices have not been detected in the checkout procedure described, serious damage and/or personal injury can result when the power is turned on.

In order to minimize the risk of damage and/or injury, there should be no load on the motor control center when it is energized. Turn off all of the downstream loads, including distribution equipment and other devices which are remote from the motor control center.

The equipment should be energized in sequence by starting at the *source end* of the system and working toward the *load end*. In other words, energize the main devices, then the feeder devices, and then the branch-circuit devices. With barriers (if applicable) in place and unit doors closed and latched, turn on the devices with a firm positive motion. Protection devices that are not quick-acting should not be "teased" into the closed position.

After all disconnect devices have been closed, loads such as lighting circuits, starters, contractors, heaters, and motors can be turned on.

Padlock Provisions

The disconnect can be padlocked in the OFF position with unit door open or closed with up to three padlocks. The door cannot be opened or closed with the disconnect padlocked, and the disconnect cannot be forced closed.

Advance/ Retract Mechanism Warning

There is a hazard of electrical shock or burn whenever working in or around electrical equipment. Turn off the power supply to the equipment before working inside motor control centers. Operating motor control units with doors open or not properly secured is potentially dangerous and can result in personal injury. (See manufacturer's specifications for advancement and retraction.)

Standard Motor Circuit Protector

Circuit Breaker Application Information

The *National Electrical Code*® requires that magnetic starters, used in combination with adjustable magnetic-trip-only circuit breakers, have an overload relay in each conductor.

The adjustable magnetic trip setting will be set at LO by the factory. The user may need to adjust this setting for proper motor start-up. The 700 percent and 1300°/O of full load current (FLI) set point limits are outlined in *NEC* Table 430–152 and exceptions in Section 430–52.

After obtaining the motor FLI from the motor nameplate, the user may then select the adjustable trip setpoint of 700 percent FLI to test-start the motor. Further adjustments may be required for motor load characteristics (not to exceed 1300 percent FLI).

Fuse Clip Location

Thirty (30) and 60 amp fuse bases should be installed for the proper fuse class and maximum voltage. The base pan of the switch has five sets of mounting holes for this purpose.

Mount the lower fuse base in the proper mounting holes per the manufacturer's specifications.

Maintenance

The following maintenance procedures should be followed before energizing any new motor control center equipment. Regular maintenance should be performed at least annually or more frequently depending on service conditions or established maintenance policy. Maintenance should also be performed following any service electrical fault or unusual occurrence.

Enclosure

Examine the interior and exterior of the motor control center for signs of moisture, oil, or other foreign material. If present, eliminate the source of foreign material and clean the motor control center.

Clean the interior and exterior of the motor control center with a vacuum cleaner. Do not use compressed air, as it will only redistribute contaminants to other surfaces. Check the enclosure for any damage that might reduce electrical clearances.

Examine the finish of the enclosure. Touch up the paint if necessary. Replace any badly corroded or damaged enclosure parts.

Bus Bars and Incoming Line Compartments

WARNING: There is a hazard of electrical shock or burn whenever working in or around electrical equipment. Turn off the power supply to the equipment before working on it.

Maintenance of bus and incoming line lug connections should be performed at least annually, or more frequently depending on service conditions and your established maintenance policy. The procedure outlined below should be followed at the time of installation to locate and tighten any connections which may have loosened during shipment and handling.

1. Wire: Remove the top horizontal wire trough covers in each section.
2. Bus: Expose the bus and bus connections by removing the two-piece bus barrier in each section. Examine all bus bars and connectors. Replace any parts that are badly discolored, corroded, pitted, or have been subjected to excessive temperatures.

> **CAUTION:** Do not attempt to clean bus bars or connectors that are damaged in any way. Replace them with new parts.

> **CAUTION:** Never brush or sandpaper the bus. This will remove plating and cause oxidation. Use a cleaning fluid approved for such use. Do not use cleaning fluid on insulators.

Check, and tighten if required, all bolts at the bus connection points indicated by a hexagon in the adjacent figures. Although one specific type of compartment or bus is shown in these figures, it is intended that similar maintenance be performed on all bolted connections. Maximum torque values should be used (see manufacturer's specifications).

If a torque wrench is not available and connection requires tightening, carefully tighten the bolt until the Belleville conical washer appears flat. (At this point, a significant increase in force will be felt.)

3. Torque Range: See manufacturer's specifications.

Check, and tighten if required, all main lug set screws holding incoming conductors at main lugs. Recommended torque values are listed in *Electrician's Technical Reference: Motors,* a companion book to this text. Torque all lug set screws. Fasten all set screws holding conductors with a torque wrench. (See manufacturer's specifications for torque requirements.)

4. Insulators, Braces, and Barriers: Inspect and replace any that show signs of arcing damage, tracking, excessive heat, or cracking.
5. Control Units:
   (a) For removal, see manufacturer's specifications.
   (b) Circuit breakers and fusible disconnects
       1. Check switch blades for arching or excessive heat—replace.
       2. Check mechanical operation for proper operation.
       3. Test for proper trip, on, off, and reset indicators.
       4. Check for proper door lock interlock operations.

Insulation Test

> **WARNING:** There is a hazard of electrical shock or burn whenever working on or around electrical equipment. Turn off the power supply to the equipment before working inside motor control centers.

> **CAUTION:** Do not megger solid state devices, capacitor units, or any devices not designed to withstand megger voltage. Disconnect these devices before testing the rest of the motor control center.

Before a motor control center is placed into service (after installation or regular maintenance), resistance measurements should be taken. Use an insulation tester (megger) with a potential of 500–1000 volts.

Readings should be taken between each phase and from each phase to ground both with the branch disconnects OFF and ON. The main disconnect should be OFF during all megger tests.

Megger readings with all the disconnects OFF will typically be 5–20 megohms. Lower readings may be observed during start-up on new equipment that has been stored in a damp area. If the readings are above one megohm, a few branch units may be energized to help dry out the motor control center. If additional readings are above one megohm, additional units can be energized. After the equipment has been in operation for 48 hours, readings should be in the 5-20 megohm range.

When megger readings are taken with the disconnects ON (except for the main), all devices completing circuits between phases or between phases and neutral (e.g., control transformers) must be disconnected. Readings may be slightly lower, but the one megohm lower limit (during start-up) should be observed.

Record all megger readings below. Any sudden change in resistance values (even if within acceptable range) may indicate potential insulation failure. Early detection and replacement of faulty insulating components can help to avoid untimely failure.

If megger readings are below 5 megohms (one megohm during start-up) consult your manufacturer.

Insulation Test Checklist Procedure

Check with all disconnects open:

Date
Phase-to-Phase
a–b
b–c
c–a

Phase-to-Ground
a–gnd
b–gnd
c–gnd

Return motor control center to service. Energize main first, then each branch unit one at a time.

Maintenance after a Fault Has Occurred

WARNING: All equipment must be deenergized, disconnected, and isolated to prevent accidental contact with live parts. Check voltage on all control terminals and on all line and load terminals of circuit breakers, disconnect switches, and starters or contractors before working on this equipment. Only qualified personnel should be involved in the inspection and repair procedure, and all safety procedures must be observed.

The *excessive currents* occurring during a fault may result in damage to the structure component and/or bus and conductor damage due to mechanical distortion, thermal damage, metal deposits, or smoke. After a fault, correct the cause of the fault, inspect all equipment, and make any necessary repairs or replacements before putting the equipment into service again. Be sure all replacement parts are of the proper rating and are suitable for the application. If in doubt, consult your manufacturer.

The complete maintenance procedure (described earlier in the Maintenance section) should be followed after any fault. The following are additional items to watch for in these circumstances.

Locate and correct the cause of the fault. If the fault occurred downstream from the motor control center, appropriate maintenance should be carried out on all equipment involved.

External evidence of *enclosure* damage usually indicates damage within the unit. Extensive damage will require replacement of the enclosure parts and the enclosed equipment. Follow proper maintenance procedures. Replace any parts that are damaged or deformed. Pay particular attention to door hinges and door closing hardware. Inspect the area around any damaged units (both inside and out) for displaced parts from the damaged unit.

For *bus bars and incoming line compartments,* follow the maintenance procedures described earlier for tightening all electrical connections to their proper torques. Replace any deformed bus bars or connectors as well as any showing signs of arcing damage. Inspect for cracked and/or burned insulators.

The following procedures refer to *units:*

1. Examine the disconnect means for evidence of possible damage. See that the operator mechanism properly turns ON and OFF the disconnect. Exercise the push-to-trip feature on circuit breakers and be sure that the operator properly resets the breaker.

2. Check that the door interlock prevents the opening of the unit door while the disconnect is in the ON position.

3. Check motor starters for any damage. Replace contacts and contact springs if the contacts are welded or show heat damage. If deterioration extends beyond the contacts, replace the entire contactor or starter.

4. Replace the complete overload relay if the thermal units have been burned out or if there are any indications of arcing or burning on the relay.

5. Inspect all fuses and fuse clips. Always replace all fuses in a set even if only one or two are open circuited.

6. Check all conductors and other devices within the units for any signs of damage.

7. Complete an insulation test before attempting to put the motor control center back into service.

8. Complete preoperational check list procedure.

9. Reenergize the equipment following proper maintenance procedures.

Starters

2- and 3-Pole AC Magnetic Contactors and Starters, NEMA Size 00

To identify parts, refer to Figure 14–1A.

Auxiliary Contacts   All contactors are supplied with a normally open holding circuit contact as standard. Additional normally open or normally closed auxiliary contacts can be installed in the field.

Cover-Mounted Control Unit   NEMA Type 1 general purpose enclosures with slip-on or hinged covers are supplied with knockouts for field additions.

Overload Relays   A melting alloy overload relay is supplied as standard with provisions for one or three thermal units. The contact unit of the melting alloy overload relay is available with a normally open or normally closed isolated alarm contact in addition to the standard normally closed contact. The contact unit with alarm circuit contacts can be installed in the field (contact manufacturer for specific details).

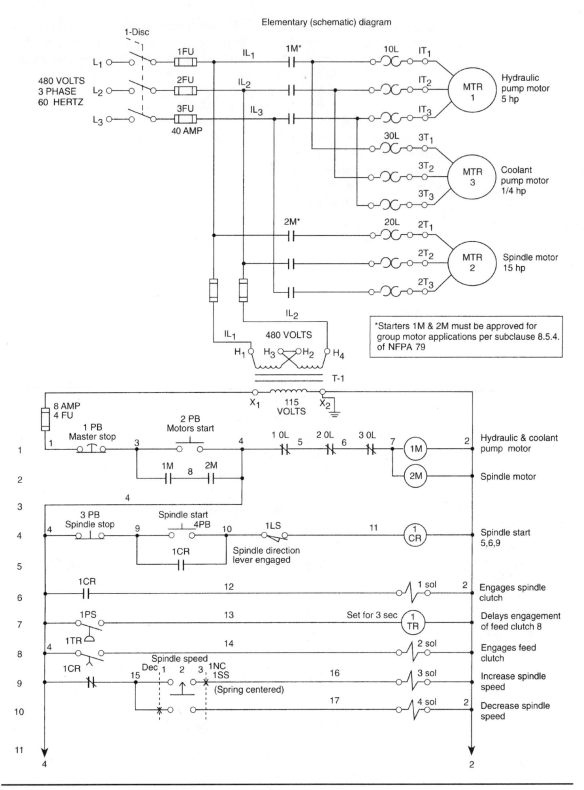

Figure 14–1A   Sample electrical diagrams.

These overload relays are not designed for field repair and should not be disassembled.

Terminals Use copper wire only on device power and control terminals. Pressure wire power terminals are suitable for wire sizes #14-8 AWG—solid or stranded. Pressure wire control terminals are suitable for wire sizes #16-12 AWG—solid or stranded.

Inspecting and Replacing Contacts Contacts are not harmed by discoloration and slight pitting. Do not file contacts, as it wastes contact material. Replacement is necessary only when the contact has worn thin. Replacement contacts for starters or contactors are available as kits from your manufacturer.

It is not necessary to remove any wiring to inspect or replace the contacts. To inspect or replace contacts, loosen the two captive screws holding the contact actuator to the contact block. Lift the contact actuator to expose the contacts.

Manual Operation Manual operation of contactors and starters may be accomplished by pushing the contact carrier down with a screwdriver. A slot suitable for this use is provided in the coil cover.

Coil Replacement To remove the coil, loosen the two captive cover screws (Figure 14–1B). Disconnect wires from the coil terminals and remove the cover. Remove and disassemble the magnet, coil, and armature unit.

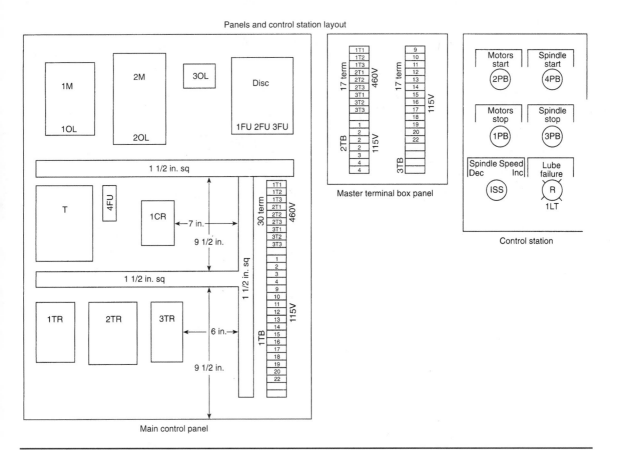

Figure 14–1B Panels and control station layout.

To replace the coil, first assemble the magnet, replacement coil, and armature. Manually operate the contact carrier and insert the complete unit. Before installing the cover, manually operate the device as described earlier in Manual Operation to ensure all parts are functioning properly. Check with the manufacturer for torque requirements.

Assembly Instructions   These instructions illustrate how contactors and starters are assembled. Factory recommended torques for mechanical, electrical, and pressure wire connection torques are also important. These torques must be followed to ensure proper device operation.

Short Circuit Protection   Short circuit protection provides branch-circuit overcurrent protection for starters, referring to instructions supplied with the thermal unit selection table. Provide branch-circuit overcurrent protection for contactors in accordance with the National Electrical Code. Do not exceed the maximum protective device ratings.

Distant Control   Series impedance and shunt capacitance of the control circuit must be considered to assure proper operation of contactors and starters when controlled from remote operator stations. Depending upon the voltage, wire size, and number of control wires used, series impedance or shunt capacitance may limit the maximum distance of the wire run.

Voltage drop calculations may be needed in some cases. The following listing gives typical distances for control wiring without voltage drop.

Maximum Control Distance

Coil Voltage Maximum	Control Distance (feet)	
(60 Hz)	#14 AWG Copper Wire	#12 AWG Copper Wire
120	1335	1800
240	540	450
480	135	110

See Appendix G for information on troubleshooting starters.

CHAPTER

15

Motors and Motor Protection

Motors

Introduction

The following is a brief description of common-use motors. Description, operation, troubleshooting, and application of motors is discussed in detail in the *Electrician's Technical Reference: Motors*.

The armature winding of a motor is made of copper wire. Because of the low resistance of the armature, short-circuit current will flow when voltage is applied. As the armature starts turning, it cuts the magnetic lines of force of the field winding which results in induced counter-electromotive force (cemf) in the armature conductor. The cemf opposes the applied line voltage, acting as an automatic current-limiter that cuts down on the armature short-circuit current. If the armature were jammed so that it could not turn, no cemf would be produced to reduce the short-circuit current. This would result in locked-rotor currents that can overheat and destroy (burn up) the motor winding.

Speed Control

One of the great advantages of the DC motor is the motor's ability to maintain a constant speed. If the speed of a DC motor increases, the armature will cut through the electromagnetic field at an increasing rate. This results in a greater armature cemf, which cuts down on the increased armature current resulting in the slowing back down of the motor.

Placing a load on a DC motor causes the motor to slow down, thereby reducing the rate at which the armature is cut by the field flux lines. As a result, the armature cemf decreases, causing an increase in the applied armature voltage and current. The increase in current results in an increase in motor speed. This gives DC motors a built-in system for regulating their own speed. For these reasons, computer disk drives and recording equipment use only DC motors.

Reversing a DC Motor

To reverse a DC motor you must reverse either the field or the armature magnetic field. This is accomplished by reversing either the field or armature current. Because most DC motors have the field and armature winding fed from the same DC power supply, reversing the polarity of the power supply will change the field and armature simultaneously. This will result in the motor continuing to run in the same direction.

AC Motors

The purest form of AC motor is the induction motor. An induction motor uses no physical connection between its rotating member rotor and stationary member stator. The most common type of three-phase motor is the AC squirrel cage induction motor which is used in almost all major industrial applications. The following describes some common types of AC motors.

Synchronous Motors   In a synchronous motor, the rotor is actually locked in step with the rotating stator field and is dragged along at the synchronous speed of the rotating magnetic field. Synchronous motors maintain their speed with a high degree of accuracy and are used for electric clocks and other timing devices.

Wound Rotor Motors   Wound rotor induction motors are used only in special applications because of their high starting torque requirements. Wound rotor induction motors only operate on three-phase AC power.

Universal Motors   Universal motors are fractional horsepower motors that operate equally well on both AC and DC power. They are used in vacuum cleaners, electric drills, mixers, and light household appliances. These motors have the inherent disadvantage associated with DC motors: the need for commutation. The disadvantage of commutation is that as the motor operates, motor parts rub against each other and the motor wears out. Pure AC motors do not depend on commutation for operation.

Reversing AC Motors

Three-phase AC motors can be reversed by reversing any two of the three line conductors. More detailed information on how to reverse a single phase motor can be found in *Electrician's Technical Reference: Motors.*

Dual Voltage Motors

Dual voltage motors are made with two field windings, each rated for low voltage. The field windings are connected in parallel for low-voltage operation and in series for high-voltage operation.

Motor Horsepower/Watts

Motors are used to convert electrical energy to mechanical work. The output mechanical work of a motor is rated in horsepower, which can be converted to electrical output of 746 watts per horsepower. Caution: this is the output horsepower conversion rating of a motor, not the input rating of the motor for calculating motor currents. The purpose of understanding motor output watts is to help you understand how motor nameplate currents are developed.

Formulas

One Horsepower = 746 watts
Horsepower = Output Watts ÷ 746
Output Watts = Horsepower × 746 watts

Example A What size horsepower motor is used to produce a 15 kW output?

Horsepower = Output Watts ÷ 746
Horsepower = 15,000 watts ÷ 746 watts
Horsepower = 20 hp

Example B What is the HP rating of a 22,380 watt output motor?

Horsepower = Output Watts ÷ 746
Horsepower = 22,380 watts ÷ 746 watts
Horsepower = 30 hp

Example C What is the electric output watts rating of a 30 HP motor?

Output Watts = HP × 746 watts
Output Watts = 30 × 746 (22,380)
Output Watts = 22,380 watts or 22.38 kW

Example D What is the output watts rating of a 10 HP, AC, three-phase, 480 volt motor, efficiency 75 percent and power factor 70 percent?

Output Watts = HP × 746 watts
Output Watts = 10 × 746 (7,460 watts)
Output Watts = 7,460 ÷ 1,000 or 7.46 kW

Efficiency, power factor, AC, three-phase, and 480 volts have nothing to do with determining the output watts of a motor.

Motor Nameplate Amperes

Motor nameplate indicates the motor operating voltage and current as well as additional factors. The motor nameplate full load current is sometimes abbreviated FLA. This FLA rating is the current the motor draws while carrying its rated horsepower load, at its rated voltage. The actual current drawn is dependent on the load on the motor: If the load increases the current also increases. It is important not to overload a motor above its rated horsepower. Caution: motors can be damaged by under voltage, which also causes the motors to operate at overload current.

Motor nameplate current is expressed in amperes and is calculated by the following steps:

Step 1

Determine the motor output watts.
Output Watts = Horsepower (HP) × 746

Step 2

Determine the motor input watts.
Input = Output ÷ Efficiency (EFF)

Step 3

Determine the motor input volt amps (VA)
VA = Watts ÷ Power Factor (PF)

Step 4

Determine the motor amperes.

Motor nameplate amperes can be determined by the following formulas:

Single-Phase

Input (I) = Motor HP × 746 watts ÷ Volts (E) × Eff × PF

Three-Phase

Input(I) = Motor HP × 746 watts ÷ E × √13 × Eff × PF

Example A What is the nameplate amperes for a 7.5 HP motor, 240V, single-phase, power factor 87 percent, efficiency 93 percent?

1. Determine the motor output watts.
   Output Watts = HP × 746 watts
   7.5 × 746 = 5,595 watts

2. Determine the motor input watts.
   Input = Output ÷ Eff
   5,595 watts ÷ .93 = 6,016 watts

3. Determine the motor input VA.
   VA = Watts ÷ PF
   6,016 watts ÷ .87 = 6,915 VA

4. Determine the motor amperes.
   I = VA(6,915) ÷ volts (240) = watts (28.81amps)

OR

Nameplate amperes = Horsepower × 746 watts ÷ E × Eff × PF
Nameplate amperes = 7.5 × 746 watts ÷ 240 × .93 × .87
Nameplate amperes = 28.81 amperes

Example B What is the nameplate amperes of a 40 HP, 208v, 3-phase motor, power factor 90 percent and efficiency 80 percent?

1. Determine the motor output watts.
   Output watts = HP × 746
   40 × 746 = 29,840 watts

2. Determine motor input watts,
   Input = Output ÷ Eff
   29,840 watts ÷ .80 = 37,300 watts

3. Determine motor input VA.
   VA = Watts ÷ PF
   37,300 ÷ .90 = 41,444 VA

4. Determine the motor amperes.
   I = VA ÷ E × √3 × Eff × PF
   41,444 ÷ 208 × 1.732 × .8 × .9 = 115 nameplate amperes

OR

Nameplate amperes = Horsepower × 746 watts ÷ E × √3 × EFF × PF
Nameplate amperes = 40 × 746 watts ÷ 208 × 1.73 × .8 × .9
Nameplate amperes = 115 amperes

Motor Protection

Motors can be damaged, or their effective life can be reduced, when subjected to a continuous current only slightly higher than their full load current rating times the service factor.

NOTE: Motors are designed to handle inrush or locked rotor currents without excessive temperature rise, provided the accelerating time is not too long nor duty cycle too frequent.

Damage to insulation and windings of the motor can also be sustained on extremely high currents of short duration, as in ground faults and short circuits.

All currents in excess of full load current can be classified as overcurrents. In general, however, a distinction is made based on the magnitude of the overcurrent and the equipment to be protected.

An overcurrent up to locked rotor current is usually the result of a mechanical overload on the motor.

Overcurrents due to short circuits or ground faults are much higher than locked rotor currents. Equipment used to protect against damage due to this type of overcurrent must not only protect the motor, but also the branch circuit conductors and the motor controller.

Motor overload protection differs from overcurrent protection, and each is separately covered in the following paragraphs.

Overcurrent Protection

The function of the overcurrent protective device is to protect the motor branch circuit conductors, control apparatus, and motor from short circuits or ground faults. The protective devices commonly used to sense and clear overcurrents are thermal magnetic circuit breakers and fuses. National Electrical Manufacturers Association (NEMA) type code letters on motor nameplates indicate the maximum percentage of full load current that the motor is allowed to endure. The short circuit device must be capable of carrying the starting current of the motor, but the device setting must not exceed 250 percent of full load current (with no NEMA type code letter on the motor), or from 150 to 250 percent of full load current (depending upon NEMA type code letter on the motor) may be increased, but must never exceed 400 percent of the motor full load current.

The *National Electrical Code® (NEC®)* requires (with a few exceptions) that there be a means to disconnect the motor and controller from the line, in addition to an overcurrent protective device to clear short circuit faults. A circuit can incorporate fault protection and disconnect in one basic device. When the overcurrent protection is provided by fuses, a disconnect switch is required and the switch and fuses are generally combined.

Motor Overload

If a motor is overloaded, the operating current increases to a value above the rated motor full-load current (FLA). The overload current may damage the motor due to excess heat. The motor nameplate current is used to size the motor overload protection device according to the *NEC®*. For all practical purposes you will not need to calculate the motor nameplate current, but you should understand how it is calculated.

A motor has no intelligence and will attempt to drive any load, even if excessive. Exclusive of inrush or locked rotor current when accelerating, the current drawn by the motor when running is proportional to the load, varying from a no-load current (approximately 40 percent of full-load current) to the full-load current rating stamped on the motor nameplate. When the load exceeds the torque rating of the motor, it draws higher than full-load current and the condition is described as an overload. The maximum overload exists under locked rotor conditions, in which the load is so excessive that the motor stalls or fails to start, and as a consequence draws continual inrush (locked rotor) current.

Overloads can be electrical or mechanical in origin. Single phasing of a polyphase motor and low line voltage are examples of electrical overloads.

Overload Protection   The effect of an overload is an excessive rise in temperature in the motor winding due to current higher than full-load current. The larger the overload, the more quickly the temperature will increase to a point that is damaging to the insulation and lubrication of the motor. An inverse relationship therefore exists between current and time: the higher the current, the shorter the time before motor damage, or burnout occurs.

All overloads shorten motor life by deteriorating the insulation. Relatively small overloads of short duration cause little damage, but if sustained they are just as harmful as overloads of greater magnitude. The ideal overload protection for a motor is an element with current sensing properties similar to the heating curve of the motor, which acts to open the motor circuit when full-load current is exceeded. The operation of the protective device should allow the motor a harmless overload, but quickly remove the motor from the line when an overload persists for too long.

Overload Protection—Fuses   Fuses are not designed to provide overload protection. Their basic function is to protect against short circuits (overcurrents). When starting, motors draw a high inrush current generally six times that of the normal full-load current. Single element fuses have no way of distinguishing between this temporary and harmless inrush current and a damaging overload. Thus, a fuse chosen on the basis of motor full-load current would blow every time the motor started. On the other hand, if a fuse were chosen large enough to pass the starting or inrush current, it would not protect the motor against small, harmful overloads which might occur later. Dual element or time delay fuses can provide motor overload protection, but they suffer the disadvantage of being nonrenewable and must be replaced.

Overload Protection—Overload Relays   The overload relay is the core of motor protection. Like the dual element fuse, the overload relay has inverse trip time characteristics, permitting it to hold in during the accelerating period when inrush current is drawn, yet providing protection on small overloads above full-load current when the motor is running. Unlike the fuse, the overload relay is renewable and can withstand repeated trip and reset cycles without need of replacement. It should be emphasized that the overload relay does not provide short circuit protection. This is the function of overcurrent protective equipment such as fuses and circuit breakers.

The overload relay consists of a current sensing unit connected in the line to the motor along with a mechanism (actuated by the sensing unit) that serves to directly or indirectly break the circuit. In a manual starter, an overload trips a mechanical latch, causing the starter contacts to open and disconnect the motor from the line. In magnetic starters, an overload opens a set of control circuit contacts within the overload relay itself. These contacts are wired in series with the starter coil in the control circuit of the magnetic starter. Breaking the coil circuit causes the starter power contacts to open, disconnecting the motor from the line.

Alarm Contacts   Standard overload relay contacts are closed under normal conditions and open when the relay trips. An alarm signal is sometimes required to indicate when a motor has stopped due to an overload trip. This is done by fitting the overload relay with a set of contacts which close when the relay trips, completing the alarm circuit. These contacts are called alarm contacts and may be used for applications such as indicating lights and audible alarms.

Magnetic Overload Relay   A magnetic overload relay has a movable magnetic core inside a coil which carries the motor current. The magnetic flux set up inside the coil pulls the core upwards. When the core rises far enough (determined by the current and the position of the core) it trips a set of contacts on the top of the relay. The movement of the core is slowed by a piston working in an oil-filled dash pot (similar to a shock absorber) mounted below the coil. This produces an inverse-time characteristic.

The effective tripping current is adjusted by moving the core on a threaded rod. The tripping time is varied by uncovering oil bypass holes in the piston. Because of the time and current adjustments, the magnetic overload relay is sometimes used to protect motors having long accelerating times or unusual duty cycles. The instantaneous trip magnetic overload relay is similar, but has no oil-filled dash pot.

Thermal Overload Relay   As the name implies, thermal overload relays react to the rising temperatures caused by the overload current to trip the overload mechanism. Thermal overload relays can be further subdivided into two types: melting alloy and bimetallic.

Melting Alloy Thermal Overload Relay   In melting alloy thermal overload relays, the motor current passes through a small heater winding. Under overload conditions, the heat causes a special solder to melt, allowing a ratchet wheel to spin free thus opening the control circuit contacts. When this occurs, the relay is said to "trip." To obtain appropriate tripping current for motors of different sizes, or different full-load currents a range of thermal units (heaters) is available. The heater coil and solder pot are combined in a one piece, nontamperable unit. Melting alloy thermal overload relays must be reset by a deliberate hand operation after they trip. A reset button is usually mounted on the cover of enclosed starters. Thermal units are rated in amperes and are selected on the basis of motor full-load current, not horsepower.

Bimetallic Thermal Overload Relay   Bimetallic thermal overload relays employ a U-shape bimetal strip associated with a current-carrying heater coil. When an overload occurs, the heat will cause the bimetal to deflect and operate a control circuit contact. Different heaters give different trip points. In addition, most relays are adjustable over a range of 85 percent to 115 percent of the nominal heater rating.

These relays are field convertible from hand reset to automatic reset and vice versa. On automatic reset after tripping the relay, contacts will automatically reclose when the relay has cooled down. This is an advantage when the relays are inaccessible. However, automatic reset overload relays should not normally be used with 2-wire control. With this arrangement, the motor will restart when the overload relay contacts reclose after an overload relay trip, and unless the cause of the overload has been removed, the overload relay will trip again. This cycle will repeat itself and eventually the motor will burn out due to the accumulated heat from the repeated inrush current. More important is the possibility of danger to personnel. The unexpected restarting of a machine may find the operator or maintenance technician in a hazardous situation as she attempts to find out why the machine has stopped.

Ambient Compensation   Ambient-compensated bimetallic overload relays are designed for one particular situation: when the motor is at a constant ambient temperature and the controller is located separately in a varying ambient temperature. In this case, if a standard thermal overload relay were used, it would not trip at the same level of motor current if the controller temperature changed. The standard thermal overload relay is always affected by the surrounding temperature. To compensate for temperature variations, an ambient-compensated overload relay is used. Its trip point is not affected by temperature, and it performs consistently at the same value of current.

Thermal Overload Trip Relay Characteristics Melting alloy and bimetallic overload relays are designed to approximate the heat actually generated in the motor. As the motor temperature increases, so does the temperature of the thermal unit. The motor and relay heating curves shown in Figure 15–1 illustrate this relationship. We see that no matter how high the current is drawn, the overload relay will provide protection yet will not trip unnecessarily.

Thermal Overload Relay Selection Motor full-load current, the type of motor, and the possible difference in ambient temperature between the motor and the controller must all be considered when choosing overload relay thermal units or overload heaters. Motors of the same horsepower and speed do not all have the same full-load current. Always refer to the motor nameplate for the full-load current. Do not use a published table because these tables of motor full-load currents show the average or normal full-load currents, whereas the full-load current of the motor in question may be quite different. Thermal unit selection tables are published on the basis of continuous duty motors with 1.15 service factors operating under normal conditions. The tables are shown in the manufacturer's specifications and also appear on the inside of the door or cover of the controller. These selections will properly protect the motor and enable it to develop its full horsepower, allowing for the service factor if the ambient temperature is the same at the motor as at the controller. If the temperatures are not the same, or if the motor service factor is less than 1.15, a special procedure is required to select the proper thermal unit.

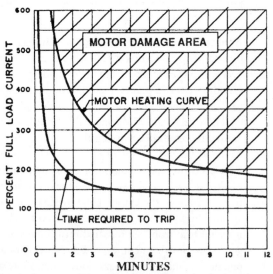

Graph shows motor heating curve and overload relay trip curve.
Overload relay will always trip at a safe value

Figure 15–1 Overload relay trip curve. Courtesy Square D Company.

Motor Starting Current

When a motor starts, the current drawn is approximately six times that of the motor full-load amperes (running current). This is also known as motor *locked-rotor amperes (LRA)*. When a motor first starts, the current is at locked-rotor current. This is because there is no initial cemf. Generally motors can withstand inrush current of six times the motor's rated FLA for short periods of time without damage to the motor windings.

If the motor operates at LRA for an extended period of time, the motor insulation and winding can be destroyed by excessive heat. Most motors can operate at 600 percent of motor FLA satisfactorily for a period of less than one minute or at 300 percent of the motor FLA for not more than three minutes.

When sizing conductors and overcurrent protection for motors, we must comply with the requirements of *NEC®* Article 430, not Article 240 [240–3(f)]. The motor full-load currents listed in Tables 430–147 through 430–150 must be used (instead of the motor nameplate rating) when sizing conductors [430–22(a)] and short-circuit ground-fault protection devices [430–52]. The motor nameplate current rating must be used to size the motor overload protection device [430–32 and 430–341].

Branch Circuit Conductors

Branch circuit conductors to a single motor must have an ampacity of not less than 125 percent of the motor full-load current as listed in Tables 430–147 through 430–150 [430–6(a) and 430–22(a)]. The actual conductor size must be selected from Table 310–16 according to the terminal temperature rating (60° or 75° C) of the equipment [110–14(c)].

Example What size THHN conductor is required for a 2 horsepower 240 volt motor? Table 430–148 lists the FLC for a 2 horsepower motor as 12 ampere, and $12 \times 1.25 = 15$ amperes (Table 310–16). Therefore, *No. 14 THHN rated 20 amperes @ 60° C* is the required conductor.

Branch Circuit Overcurrent Protection Motors and their associated equipment must be protected against overcurrent (over load, short-circuit, or ground-fault), but because of the special characteristics of induction motors, overcurrent protection is generally accomplished by having the overload protection separate from the short-circuit and ground-fault protection Part C of Article 430 contains the requirements of overload protection and Part D of Article 430 contains the requirements of short-circuit and ground-fault protection.

Overload is the condition where current is greater than the equipment ampacity rating resulting in equipment damage due to dangerous overheating (Article 100). Overload protection devices, sometimes called heaters, are intended to protect the motor, the motor-control equipment, and branch-circuit conductors from excessive heating due to motor overload [430–31], but not from short-circuits or ground-fault currents. Overload protection can be accomplished by the use of fuses, if a fuse is installed in each ungrounded conductor [430–36] and [430–5]. The branch-circuit, short-circuit, and ground-fault protection device is intended to protect the motor, the motor control apparatus, and the conductors against overcurrent due to short-circuit or ground-fault, but not against an overload [430–51].

NOTE: This type of ground-fault protection is not the type required for personnel [210–8], feeders [215–9 and 240–13] services [230–95], or construction sites receptacles [305–6(a)].

Overload Protection [430–32 and 34]

The *NEC®* requires motor overloads to be sized according to the motor nameplate current rating (FLA), not the motor full-load current rating (FLC). The standard size overload must be sized according to the requirements of Section 430–32 of the *NEC®*. If the overload sized according to Section 430–32 is not capable of carrying the motor starting and running current, the next size up overload can be used, if sized according to the requirements of Section 430–34.

Motors rated more than 1 horsepower without integral thermal protection [430–32(2)] and motors rated 1 horsepower or less automatically started [430–32(c)], must have an overload device sized in response to the motor *nameplate current rating* [430–6(a)]. The sizing of the overload is dependent on the following nameplate markings:

Service Factor   Motors with a nameplate service factor (SF) rating of 1.15 or more must have the overload protection device sized to no more than 125 percent of the motor nameplate current rating. If the overload (sized at 125 percent) is not capable of carrying the motor starting or running current, the next size up protection device is permitted, but it cannot exceed 140 percent of the motor nameplate current rating.

NOTE: Service factor of 1.15 means that the motor is designed to be loaded to 115 percent of its rated horsepower continuously. Because the motor is designed to operate at a higher capacity (load), the running current of the motor will be greater and the overload must be adjusted to permit the increased current flow.

Temperature Rise   Motors with a nameplate temperature rise rating not over 40° C must have the overload protection device sized at no more than 125 percent of motor nameplate current rating. If the overload (sized at 125 percent) is not capable of carrying the motor starting or running current of the motor, the next size up is permitted, but it cannot exceed 140 percent of the motor nameplate current rating.

NOTE: Nameplate temperature rise not over 40° C means that the motor is designed to operate so that it will not heat up over 40° C. Because the motor is designed to run cooler, the motor can carry more current.

All Other Motors   Motors that do not have a service factor of 1.15 and up, or a temperature rise rating over 40° C, must have the overload protection device sized at not more than 115 percent of the motor nameplate ampere rating. If the overload (sized at 115 percent) is not capable of carrying the motor starting or running current, the next size up device is permitted, but it cannot exceed 130 of the motor nameplate current rating.

NOTE: Overloads are available with longer time-delay characteristics, which can be used to avoid the need for using oversized overloads [430–34 FPN].

Section 430–34 states that the motor full-load current is used to size overloads. It should state a motor's nameplate full-load current for sizing overloads.

Example A   If a dual-element fuse is used for overload protection, what size fuse is required for a 5 horsepower, 240 volt, single-phase motor, service factor 1.16, motor nameplate current rating of 28 amperes (FLA)?

Answer: A 35 ampere dual-element fuse is required because overloads are sized according to the nameplate current rating, not the motor full-load current [430–32(a)(1) and 430–34]. Standard 28 × 1.25 = 35 amperes [240–6]. Maximum 28 × 1.4 = 39.2 amperes.

Example B   If a dual-element fuse is used for the overload protection, what size fuse is required for a 5 horsepower, 208 volt, single-phase motor, temperature rise 39° C, motor nameplate current rating of 28.8 amperes (FLA)?

Answer: A 35 ampere fuse is required. $28.8 \times 1.25 = 36$ amperes; next size down is 35 amperes [430–32(a)(1) and 240–6]. If the 35 ampere fuse blows, then a 40 ampere fuse would be permitted. $28.8 \times 1.4 = 40$ amperes [430–34].

Number of Overloads [430–37]

As a general rule, an overload must be installed in each ungrounded conductor according to the requirements of Table 430–37. If a fuse is used for overload protection, then a fuse must be installed in each ungrounded conductor [430–36].

Cord- and Plug-Connected Equipment [430–42]

Cord- and plug-connected motors 1 horsepower or more must have an overload device that is integral with the motor [430–42(b)]. Overload protection is not required for cord- and plug-connected motors not over 1 horsepower [430–42(c)].

Overload Alarm [430–44]

If orderly shutdown is necessary to reduce hazards to persons, the overload sensing device can be connected to a supervised alarm instead of causing the motor to shutdown.

Branch Circuit Short-Circuit Ground-Fault Protection [430–52]

In addition to overload protection, each motor and its accessories requires short-circuit and ground-fault protection according to the requirements of Section 430–52. When sizing the branch circuit protection device the following factors must be considered:

1. The motor type, such as induction, synchronous, wound-rotor, etc.
2. The motor code letter starting characteristics.
3. The type of protection device to be used, the fuse, or breaker.

National Electric Code® Section 430–52 requires the motor branch circuit short-circuit and ground-fault protection (except torque motors) to be sized no greater than the percentages listed in Table 430–152. When the short-circuit ground-fault protection device value determined from Table 430–152 does not correspond with the standard rating of overcurrent protection devices as listed in Section 240–6, the next smaller device size must be used.

Example   If the dual-element fuse selected (sized not greater than the percentages listed in Table 430–152,) is not capable of carrying the load, the next larger sized dual-element fuse can be used. The next size dual-element fuse cannot exceed 225 percent of the motor full-load current rating [430–52(a), Exception No. 2].

Inverse Time Circuit-Breaker   If the inverse time breaker selected (sized not greater than the percentages listed Table 430–152) is not capable of carrying the load, the next larger sized inverse time breaker can be used. The next size inverse time breaker cannot exceed 400 percent of the motor full-load current rating for inverse time breakers exceeding 100 amperes, or 300 percent of the motor full-load current rating for over 100 amp breakers [430–52 (a), Exception 2c].

> **WARNING:** Conductors are sized at 125 percent of the motor full-load currents [430–6(a) and 430–22(a)], overloads are sized from 115 percent to 140 percent of the motor nameplate amperes; and the motor short-circuit ground-fault protection device is sized from 150 percent up to 400 percent of the full-load current. As you can see, there is no relationship between the branch circuit conductor ampacity (125 percent), the overload size (115 percent to 140 percent), or the short-circuit ground-fault protection device (150 percent to 400 percent).

Single Branch Circuit Overcurrent Protection Device

If remote control of the motor is not necessary, considerable savings in installation cost can be achieved by using dual-element fuses as a single unit for overcurrent protection (overload load, short-circuit and ground-fault protection). The dual-element fuse must be sized according to the requirement of Section 430–32 and must be installed in series with each conductor [430–55].

Feeder Conductor Sizing [Section 430–24]

Motors Only   Feeder conductors that supply several motors must have an ampacity of not less than: (1) 125 percent of the highest rated motor full-load current (FLC) [430–17], plus (2) the sum of the full-load currents of the other motors (on the same line and phase) [430–6(a)]. For example: What size feeder conductor (in amperes) is required for two motors: Motor 1 FLC is 40 amperes and Motor 2 FLC is 30 amperes?

Answer: An 80 ampere feeder conductor is required. $(40 \times 1.25) + 30 = 80$ amperes.

Motors with Other Loads   Feeder conductors that supply motors and other loads must have an ampacity of not less than: (1) 125 percent of the highest rated motor FLC, plus (2) the sum of the full-load currents of the other motors, plus (3) the calculated load according to Article 220.

Feeder Protection [430–62(a)]: Motors Only

Motor feeder conductors (sized according to 430–24) must have a feeder protection device to protect against short-circuits and ground-faults (not overloads), sized not greater than: (1) The largest branch-circuit short-circuit ground-fault protection device [430–52] of any motor within the group, plus (2) the sum of full-load currents of the other motors (on same line and phase).

Example   What size feeder is required for a 5 horsepower, 240 volts, single-phase motor (Code letter B) and a 3 horsepower, 240 volt, single-phase motor (no code letter), given the following conditions:

    (a) The 5 horsepower motor has No. 10 THHN with a 50 ampere breaker.
    (b) The 3 horsepower motor has No. 12 THHN with a 40 ampere breaker.
    (c) The feeder conductors must have an ampacity of 52 amperes.
    (d) The feeder protection must not exceed 60 amperes.

The answers can be determined as follows:

1. 5 horsepower motor FLC = 28 amperes [Table 430–148]; 3 horsepower motor FLC = 17 amperes [Table 430–148].

2. Determine branch circuit conductor size [110–14(c), Table 310–16 and 430–22(a)].
   (a) 5 horsepower: $28 \times 1.25 = 35$ amperes and No. 8 THHN (40 amperes at 60° C)
   (b) 3 horsepower: $17 \times 1.25 = 21$ amperes and No. 12 THHN (25 amperes at 60° C)

3. Determine branch circuit protection size [240–6, 430–52 and Table 430–152].
   (a) 5 horsepower: $28 \times 2 = 56$ amperes; next size down = 50 amperes
   (b) 3 horsepower: $17 \times 2.5 = 43$ amperes; next size down = 40 amperes

4. Determine feeder conductor size [110–14(c), Table 310–16 and 430–24].
   $(28 \times 1.25) + 17 = 52$ amperes and No. 6 THHN (rated 55 amperes at 60° C)

5. If the 3 horsepower motor is running (17 amperes) and the 5 horsepower motor starts (50 amperes), the feeder protection must not be greater than 50 + 17 (67) amperes or the next size down: 60 amperes [240–6 and 430–62].

Feeder Protection with Other Loads [430–63]

Feeder conductors that supply motors and other loads must be protected against short-circuits and ground-faults by using the following methods: (1) The largest motor branch-circuit short-circuit ground-fault protection device for any motor of the group, plus (2) the sum of full-load current's of the other motors, plus (3) the calculated load as determined in Articles 210 or 220.

Feeder Protection Sizing [Section 430–62(a)]

Feeder Protection: 60 Amp IT Breaker
Feeder Conductors: No. 6 THHN rated 55A at 60° C
No. 10 THHN at 60° C
Branch circuit Protection: 50A
5 HP, 240V, 1-Phase, Code Letter B
FLC = 28
No. 12 THHN at 60° C
Branch circuit Protection: 40A

Feeder protection is sized *not greater* than the largest branch-circuit short-circuit ground-fault protection device plus the sum of the FLC of the other motor(s). For example, 50 amp (Motor 1) is the largest branch-circuit fuse or circuit breaker, and the "other" FLC(s) is 17 amperes (Motor 2). 50 + 17 = 67 amperes; next size down = 60 amp device [240–6(a)].

Highest-Rated Motor [430–17]

When selecting the feeder conductors and short-circuit ground-fault protection devices, the highest-rated motor shall have the highest-rated full-load current, not the highest horsepower rating [430–17]. In addition, when calculating the feeder conductor and protection, you may omit motors that do not operate at the same time, or motors that are not on the same phase.

Motors are connected so that only one motor can operate at a time. A control device allows only one motor to operate at one time. In this example, the feeder conductors and feeder short-circuit ground-fault protection sizes are based on only one motor. This is because only one of these motors will ever be on at any given time.

Example   Which is the highest-rated motor of the following?

    (a) 10 horsepower, 3 phase, 208 volt
    (b) 5 horsepower, 1 phase, 208 volt
    (c) 3 horsepower, 1 phase, 120 volt
    (d) Any of these

Answer: (c) 3 horsepower, 1 phase, 120 volt, FLC of 34 amperes

    10 horsepower = 30.8 amperes [Table 430–150]
    5 horsepower = 30.8 amperes [Table 430–148]
    3 horsepower = 34.0 amperes [Table 430–148]

Before doing any motor calculation, it is recommended that you produce a simple diagram of the motor with its information. See article 430–1 of the *NEC®* for a diagram that shows references that make it easier to calculate and reduces the chance of error.

Programmable Motor Protection

The programmable motor protection is a multifunction, motor-protective relay that monitors three-phase AC and makes separate trip and alarm decisions based on preprogrammed motor current and temperature conditions. This unit's motor protection algorithm is based on proven positive and negative (unbalance) sequence current sampling and true rms calculations. (An algorithm is a special set of rules or instructions that perform a specific operation.)

By programming this unit with the motor's electrical characteristics (such as full-load current and locked rotor current), the unit's algorithm will automatically tailor the optimal protection curve to the motor being monitored. No guesswork or approximation is needed in selecting a given protection curve because this unit matches the protection from an infinite family of curves to each specific motor. Application-related motor-load problems are further addressed through the use of such functions as jam, underload, and ground-fault protection.

A few of the many protective features of this unit are:

1. Instantaneous overcurrent trip level and start delay.
2. Instantaneous overcurrent disable setting.
3. Locked rotor current.
4. Maximum allowable stall time.
5. Ultimate trip current level.
6. I^2t alarm level.
7. Zero-sequence ground fault trip level with start and run delays.

Electrical Metering and Voltage Protection*

The following lists some of the features of metering and voltage protection:

1. Phase loss. Voltage phase loss occurs if less than 50 percent of the nominal line voltage is detected. Current phase loss occurs if smallest phase current is less than 1/16 of the largest phase current. (It updates itself twice per second; all other protection functions are updated once per second.)

2. Phase unbalance. An imbalance occurs if the maximum deviation between any two phases exceeds the amount of unbalance as a percent of nominal line voltage preset by DIP switches. Range: 5 percent to 40 percent (5 percent increments).

3. Phase reversal. Reversal occurs if any two phases become reversed for more than one second.

4. Over voltage. DIP switch setting as a percent of nominal line voltage. Range: 105 percent to 140 percent (5 percent increments).

5. Under voltage. DIP switch setting as a percent of nominal line voltage. Range: 95 percent to 60 percent (5 percent increments).

6. Delay. Allows existence of overvoltage, undervoltage, or voltage unbalance before an alarm or trip occurs. Range: 0 to 8 seconds (1 second increments).

Selecting Protective Devices

Important factors to consider when selecting protective devices are:

- Size (current rating)
- Is a time lag required?
- Interrupting capacity
- Ambient temperature where the device is to be located
- Voltage rating
- Number of poles
- Mounting requirements
- Type of operator
- Enclosure, if required

Voltage and Frequency Surges

Protecting electrical equipment against voltage surges caused by switching or lightning is very important. Even low surges that exceed the equipment insulation rating can cause the insulation to eventually weaken to the point of failure. For example, the insulation in motor windings can be stressed to the point that failure will result between windings and frame or between stator coils.

In case of a surge caused by lightning, the most severe damage occurs if there is a direct hit. Surge voltages can also be caused by induction on the line due to the lightning. The

*Adapted with permission from Rexford, *Electrical Control for Machines,* Delmar Publishers.

voltage from a direct hit will be twenty to thirty times that caused by induction. In these cases it can be considered a high-voltage and high-frequency pulse.

A lightning arrester will limit the crest of the surge by breaking down and conducting to ground. After grounding, the arrester then returns to its initial condition of nonconducting.

Switching voltage surges are not as important as lightning voltage surges. In most cases they will not exceed three times normal voltage. The highest overvoltage will be present when there is a ground-fault in the system.

For switching voltage surges the current-limiting fuse is used. The operating characteristic of this type of fuse is such as to interrupt in the first quarter of the cycle.

Controller Circuit and Conductor Protection

Motor Control Circuits

A *motor control circuit* is the circuit that carries the electric signals that direct the performance of the controller, but does not carry the main power current. [430–71, 73; 725–11(b), 14; 402–11 Ex.]. (See Figure 15–2.)

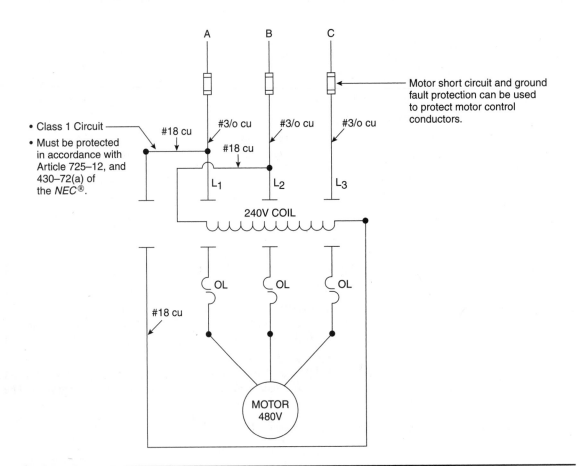

Figure 15–2 Control circuit and conductor protection.

Motor Control Circuit Conductors

Conductor Size   Two factors you should consider are:

1. When the starter coil is first energized (approximately 5 to 20 milliseconds) the coil will require many times more current than normal. This high inrush current can produce a voltage drop that can prevent the starter contacts from closing properly. The conductors should be sized large enough to prevent excessive voltage drop.

2. The control circuit conductors must be sized so that they are protected according to the NEC requirements of Section 430–72 and its exceptions. See Tables M, O, P, Q, R, S1, and S2 of the *Electrician's Technical Reference: Motors.*

Protection of Control Circuits [Section 430–72(a)]   Class 1 control circuits must be protected according to 725–12, 430–72(a).

Overcurrent Protection [430–72(a)]

Motor control conductors not tapped from the motor branch circuit protection device are generally Class 1 remote-control conductors and must be protected according to the requirements of Section 725–12. If the circuit is from a Class 1 power supply, then the conductors must be protected according to the requirements of Section 725–25.

Motor control circuit conductors tapped from the motor branch-circuit protection device must have overcurrent protection according to the requirements of Section 430–72(b), Exception No. 2 (ampacities as listed in Table 310–16). Tapped motor control conductors are not considered branch circuit conductors and can be protected by supplemental overcurrent protection devices [240–10] or by the branch circuit protection device. (See Figure 15–2.)

Improper overcurrent protection and grounding can present a potential shock hazard. Figure 15–3 shows how multiple ground faults can cause a motor to operate unintentially.

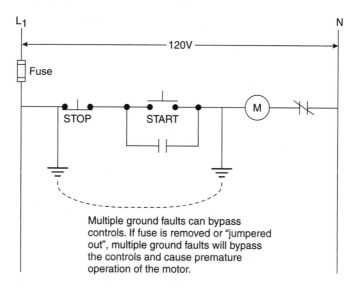

Figure 15–3   Article 430-72 of the *National Electrical Code* ® requires overcurrent protection for control circuits.

This motor could operate a press, cutter, or some other machine that could cause serious damage to equipment or individuals if not controlled properly.

Summary
1. Overcurrent Protection [430–72 (a)]
   (a) Size of OCPD needed is determined by the following:
       1. Tapped from the load side of the branch circuit
       2. Branch circuit or supplementary can be used for overcurrent protection for circuits tapped on the load side of the branch circuit device
       3. Other than tap conductors described in 1 and 2 above, can be sized by using Article 725–12 or 725–35 as applicable

2. Conductor Protection [430–72 (b)]
   (a) See Table 430–72(b) to size conductors
       1. 430–72(b) Exceptions 1, 2, 3, & 4

See Appendix G for information on troubleshooting.

CHAPTER

16

Designing Control Systems

Selection of a Controller

The motor, machine, and motor controller are interrelated and must be considered as a package when choosing a specific device for a particular application. In general, six basic factors influence the selection of a controller:

- Electrical service
- Operating characteristics of a controller
- Environment
- Power quality
- National codes and standards
- International Electrotechnical Commission (IEC)

Electrical Service

Establish whether the service is DC or AC. If AC, determine the voltage and the number of phases and frequency.

The electrical equipment shall be designed to operate correctly under full-load as well as no-load with the conditions of the nominal supply as specified below unless otherwise specified by the user. (Adopted from NFPA 79–1994)

AC Supplies

Steady State Voltage

Voltage:	0.9 . . . 1.1 of nominal voltage.
Frequency:	0.99 . . . 1.01 of nominal frequency continuously.
	0.98 . . . 1.02 short time.
	Note: The short time value may be specified by the user.
Harmonics:	Harmonic distortion not to exceed 10 percent of the total rms voltage between live conductors for the sum of the 2nd through 5th harmonic. An additional 2 percent

of the total rms voltage between live conductors for the sum of the 6th through 30th harmonic is permissible.

Voltage Unbalance: Neither the voltage of the negative (in 3 phase supplies) sequence component nor the voltage of the zero sequence component in 3-phase supplies exceeds 2 percent of the positive sequence component.

Voltage Impulses: Not to exceed 1.5 ms in duration with a rise/fall time between 500 ns and 500 p and a peak value of not more than 200 percent of the rated rms supply voltage value.

Voltage Interruption: Supply interrupted or at zero voltage for not more than 3 ms at any random time in the supply cycle. There shall be more than 1 second between successive interruptions.

Voltage Dips: Voltage dips shall not exceed 20 percent of the peak voltage of the supply for more than one cycle. There shall be more than 1 second between successive dips.

DC Supplies

From Batteries

Voltage: 0.85 . . . 1.15 of nominal voltage; 0.7 . . . 1.2 of nominal voltage in the case of battery-operated vehicles. Voltage interruption Not exceeding 5 ms.

From Converting Equipment

Voltage: 0.9 . . . 1.1 of nominal voltage.

Voltage interruption: Not exceeding 20 ms. There shall be more than 1 sec between successive interruptions.

Ripple (peak to peak): Does not exceed 0.05 of nominal voltage.

(See Chapter 13 for more details.)

Motor   The motor should be matched to the electrical service and correctly sized for the machine load (horsepower rating). Other considerations include motor speed and torque. To select proper protection for the motor, its full-load current rating, service factor, and time rating must be known. (See Chapter 13 for more details.)

Operating Characteristics of a Controller

The fundamental job of a motor controller is to start and stop the motor and to protect the motor, machine, and operator. The controller might also provide supplementary functions that include:

- Reversing
- Jogging or inching
- Plugging
- Operation at several speeds
- Operation at reduced levels of current and motor torque

(See Chapter 2 for more details.)

Environment

Controller enclosures serve to provide protection for operating personnel by preventing accidental contact with live parts. In certain applications, the controller itself must be protected from a variety of environmental conditions which might include:

- Water, rain, snow, or sleet
- Dirt or noncombustible dust
- Cutting oils, coolants, or lubricants

Both personnel and property require protection in environments made hazardous by the presence of explosive gases or combustible dusts. (See Chapter 1 for more details.)

Power Quality

Power quality considerations are critical to the proper operation of a motor control system. Assurance of proper application of the premise wiring system removes that as a possible problem when malfunction of equipment occurs. Consider the following:

- Grounding: Power and sensitive electronic equipment
- Shielding: High and low voltage systems
- High and low voltage transients
- Switching transients
- Harmonics
- Wiring methods
- Overcurrent protection
- Ampacity

National Codes and Standards

Motor control equipment is designed to meet the provisions of the *National Electrical Code®*. Code sections applying to industrial control devices are Article 430 on motors and motor controllers and Article 500 on hazardous locations.

The *National Electrical Code®* deals with the installation of equipment and is primarily concerned with safety (the prevention of injury and fire hazard to persons and property arising from the use of electricity). It is adopted on a local basis, sometimes incorporating minor changes or interpretations. *National Electrical Code®* rules and provisions are enforced by governmental bodies exercising legal jurisdiction over electrical installations and used by insurance inspectors. Therefore, minimum safety standards are assured.

The 1970 *Occupational Safety and Health Act,* as amended in 1972, requires that each employer furnish employment free from recognized hazards likely to cause serious harm. Provisions of the act are strictly enforced by inspection.

Standards established by the *National Electrical Manufacturers Association (NEMA)* assist users in the proper selection of control equipment. NEMA standards provide practical information concerning construction, test, performance, and manufacture of motor control devices such as starters, relays, and contactors.

One of the organizations that actually tests for conformity to national codes and standards is *Underwriters Laboratories (UL)*. Equipment tested and approved by UL is listed in an annual publication, which is kept current by means of periodic supplements reflecting the latest additions and deletions.

International Electrotechnical Commission (IEC)

The *International Electrotechnical Commission (IEC),* founded in 1906, is responsible for international standardization in the electrical and electronics fields. The IEC is presently composed of forty-one national committees, including that of the United States, that represent about 80 percent of the world's population. For motor control, the IEC issues recommendations on electrical terms, ratings, test methods, and dimensional requirements. (See Chapter 13 for more details.)

Design Considerations

The design of control systems is critical to the success and operation of the control system. Considerations such as power quality, safety, cost, manufacturing, assembly, and operating conditions enter into the final machine. The following should be considered when designing a new control system:

- The maintenance considerations of the system
- Economics in building a control system
- Environment where the control system is to be used
- Installation of electrical components with other equipment, such as mechanical, hydraulic, pneumatic, and structural
- Vibration, shock, or other mechanical motion that may affect operation of electrical components
- Workmanship that requires the least amount of maintenance possible
- Identification of all conductors and terminals, with numbers and letters corresponding to filed applications
- Terminal block spaces to accommodate future expansion and damaged terminals
- Design that allows proper wire bending space at terminal connections [See NFPA 70 *National Electrical Code*® Article 370, 373 & 374]
- Properly sized conductors and conduit. Do not leave this up to the installer to determine. Improper application of wiring methods is one of the major causes of malfunctioning of the control system and your system is only as reliable as its weakest point. (See NFPA 70 *National Electrical Code*® Chapter 9 Table 11)
- Spare conductors when the possibility of future expansion exists
- Consideration of total power requirements, including voltage drop and power factor problems. (See Chapter 13 and the *Electrician's Technical Reference: Motors)*
- Location and type of enclosures to be used
- Location of service equipment and utility-owned transformer or substation
- Location of all disconnecting means
- User's power voltage, phase, and frequency
- The type of system available (wye or delta)
- The type of grounding system needed
- Special conditions or limitations on the power system

- Ambient temperature condition
- Atmospheric conditions
- Manufacturer's specification of components
- User's specification on motor starters, if not full voltage
- A complete and accurate sequence of functions the control system is to follow. This item is probably the most difficult to obtain, due to (1) inexperience, particularly in the case of a new process, and (2) a general lack of knowledge.
- Communicate with the installer. Do not require the installer to guess at the designer's intentions.

Practical help in designing a motor control system:

1. Diagrams and layouts.
2. Locating, assembling, and installing of components.

Schematics and Wiring Diagrams

For more information on schematics and wiring diagrams, refer to the example in Figure 14-1A and Figure 14-B. Also, see Chapter 10 for more details.

Troubleshooting

Analyzing the Problem

There are predominately two methods of trouble shooting motor control systems.

The first method is called the *shotgun method* because the electrician goes to several places in the system rather than starting at the beginning of the system. This method is usually used by the electrician who is familiar with the system with which he is dealing. Sometimes this is called the "hit or miss" method. To the inexperienced electrician, this method generally creates more expense and wastes time.

The second method is called the *analysis method.* To analyze any problem, it helps to break it down into types of sections. Breaking down a problem limits the size of the job. In troubleshooting, the following causal areas might be considered:

- Electrical/Electronic
- Mechanical
- Fluid power
- Pneumatic
- Personnel

In many cases, the problem may be a combination of two or more of these areas.

The following examples illustrate how breaking down a problem into types of causes can simplify the troubleshooting procedure:

1. From advance information supplied by a machine operator or from early examination by an electrician, what first appears to be an electrical problem may turn out to be a mechanical one.

2. The failure of a limit switch to function properly may be caused by problems in the electrical contacts or the mechanical operator. However, the result of a cycle failing to complete is the same, regardless of which of these two items caused the problem.

Trouble-shooting Motor Control Systems

WARNING: There is a hazard of electrical shock or burn when working on or near electrical equipment. Turn off the power supply to the equipment before working inside motor control centers.

Troubleshooting Motor Control

This list is of a general nature and covers only the main causes of trouble. Misapplication of a device can be a cause of serious trouble. However, rather than list this cause repeatedly it should be noted here that misapplication is a major cause

1. Contacts: Contact Chatter (also see #8 "Noisy Magnet")

    Cause: 1. Poor contact in control circuit.
    2. Low voltage.

    Remedy: 1. Replace the contact device or use holding circuit interlock (3 wire control).
    2. Check coil terminal voltage and voltage dips during starting.

2. Contacts: Welding or Freezing

    Cause: 1. Abnormal inrush of current.
    2. Rapid jogging.
    3. Insufficient tip pressure.
    4. Low voltage preventing magnet from sealing.
    5. Foreign matter preventing contacts from closing.
    6. Short circuit or ground fault.

    Remedy: 1. Check for grounds, shorts, or excessive motor load current, or use larger contactor.
    2. Install larger device rated for jogging service.
    3. Replace contacts and springs, check contact carrier for deformation or damage.
    4. Check coil terminal voltage and voltage dips during starting.
    5. Clean contacts with Freon. Contacts, starters, and control accessories used with very small current or low voltage should be cleaned with Freon.
    6. Remove fault and check to be sure fuse or breaker size is correct.

3. Contacts: Short Tip Life or Overheating of Tips

    Cause: 1. Filing or dressing.
    2. Interrupting excessively high currents.
    3. Excessive jogging.
    4. Weak tip pressure.
    5. Dirt or foreign matter on contact surface.
    6. Short circuits or ground-fault.
    7. Loose connection in power circuit.
    8. Sustained overload.

    Remedy: 1. Do not file silver tips. Rough spots or discoloration will not harm tips or impair their efficiency.
    2. Install larger device or check for grounds, shorts, or excessive motor currents.
    3. Install larger device rated for jogging service.

4. Replace contacts and springs, check contact carrier for deformation or damage.
5. Clean contacts with freon. Take steps to reduce entry of foreign matter into enclosure.
6. Remove fault and check to be sure fuse or breaker size is correct.
7. Clear and tighten.
8. Check for excessive motor-load current or install larger device.

4. Coils: Open Circuit

Cause: 1. Mechanical damage.

Remedy: 1. Handle and store coils carefully.

5. Coils: Overheated Coil

Cause: 1. Over voltage or high ambient temperature.
2. Incorrect coil.
3. Shorted turns caused by mechanical damage or corrosion.
4. Under voltage, failure of magnet to seal in.
5. Dirt or rust on pole faces.
6. Mechanical obstruction.

Remedy: 1. Check coil terminal voltage, which should not exceed 110 percent of coil rating.
2. Install correct coil.
3. Replace coil.
4. Check coil terminal voltage, which should be at least 85 percent of coil rating.
5. Clean pole faces.
6. With power OFF, check for free movement of contact and armature assembly.

6. Overload Relays: Tripping

Cause: 1. Sustained overload.
2. Loose or corroded connection in power circuit.
3. Incorrect thermal units.
4. Excessive coil voltage.

Remedy: 1. Check for excessive motor currents or current unbalance, and correct cause.
2. Clean and tighten.
3. Thermal units should be replaced with correct size for the application conditions.
4. Voltage should not exceed 110 percent of coil rating.

7. Overload Relays: Failure to Trip

Cause: 1. Incorrect thermal units.
2. Mechanical binding, dirt, corrosion, etc.
3. Relay previously damaged by short circuit.
4. Relay contact welded or not in series with contactor coil.

Remedy: 1. Check thermal unit selection table. Apply proper thermal units.
2. Replace relay and thermal units.
3. Check circuit for a fault and correct condition.
4. Replace contact or entire relay as necessary.

8. Magnetic and Mechanical Parts: Noisy Magnet

Cause: 1. Broken shading coil.
2. Dirt or rust on magnetic faces.
3. Low voltage.

Remedy: 1. Replace magnet and armature.
2. Clean.
3. Check coil terminal voltage and voltage dips during starting.

9. Magnetic and Mechanical Parts: Failure to Pick Up and Seal

Cause: 1. No control voltage.
2. Low voltage.
3. Mechanical obstruction.
4. Coil open or overheated.
5. Wrong coil.

Remedy: 1. Check and control circuit for loose connection or poor continuity of contacts.
2. Check coil terminal voltage and voltage dips during starting.
3. With power OFF, check for free movement of contact and armature assembly.
4. Replace.

10. Magnetic and Mechanical Parts: Failure to Drop Out

Cause: 1. Gummy substance on pole faces.
2. Voltage not removed.
3. Worn or corroded parts causing binding.
4. Residual magnetism due to lack of air gap in magnet path.
5. Contacts welded.

Remedy: 1. Clean pole faces.
2. Check coil terminal voltage and control circuit.
3. Replace parts.
4. Replace magnet and armature.
5. See "Contacts: Welding or Freezing."

11. Pneumatic Timers: Erratic Timing

Cause: 1. Foreign matter in valve.

Remedy: 1. Replace complete timing head or return timer to factory for repair and adjustment.

12. Pneumatic Timers: Contacts Do Not Operate

Cause: 1. Maladjustment of actuating screw.
2. Worn or broken parts in snap switch.

Remedy: 1. Adjust per manufactures specifications.
2. Replace snap switch.

13. Limit Switches: Broken Parts

Cause: 1. Overtravel of actuator.

Remedy: 1. Use resilient actuator or operate within the tolerance of the device.

(See Troubleshooting Components in the following section.)

14. Manual Starters: Failure to Reset

      Cause: 1. Latching mechanism worn or broken.

    Remedy: 1. Replace starter.

(See Appendix G.)

Troubleshooting Components

Limit Switches

- Environment

    Problem: The environment causes a consistent maintenance problem.
    Solution: Relocate switch if possible. Protect the switch with a different type enclosure
    If ambient temperature is a concern, provide a thermal barrier. Shock or vibration
    can be controlled by redirecting the force with some type of absorption method.

- Actuation

    Problem: Overtravel, direction of travel, frequency of movement, and force.
    Solution: See manufacturer's specifications that come with the switch to ensure
    that the position of the cam is correct.

- Contacts

    Problem: Contacts over fail.
    Solution: Check to see if the actual current through the contacts and the switch
    current rating match.

(See Chapter 7 for more detail.)

Vane Switches   Check all the problems and solutions in the Limit Switches section and add the following:

    Problem: Accuracy of the switch does not work properly.
    Solution: Check with manufacturer's specifications to see if the vane switch is
    properly matched with the operation intended.

    NOTE: Contacts current rating on the vane switch are considerably lower than the limit switch. Typical is .75 make and .2 continuous.
    (See Chapter 7 for more detail.)

Proximity Switches   Problems with the proximity switch are similar to those of the vane switch. However, the proximity switch operates much better in adverse environments than do limit and vane switches. (See Chapter 7 for more detail.)

Push-buttons   See Chapter 7 for more detail.

Motor Control Center   See Chapter 14 for more detail.

Motor Drive / Inverter   See Chapter 12 for more detail.

Contacts and Relays   See Chapter 6 for more detail.

Starters   See Chapter 5 for more detail.

Steps to Calculate Power and Motor Control Circuits

Power

Step 1

Size branch circuit conductors

Multiply FLA of the motor by 125 percent based on Article 430–22 of the *NEC*®.
Size branch circuit conductors—Article 310–16, based on 750 C column.

Step 2

Size feeder circuit conductors

Article 430–24, 310–16

Ampacity of the conductors is based on Article 430–24. Conductor ampacity shall be the FLA sum of all motors plus (+) 25 percent of the highest rated motor in the group. Conductor size is based on 75° C, Table 310–16.

Step 3

Size overload protection

Article 430–32(a)(1), 430–6(a)

Size overload protection based on name plate FLA, temperature rise (40° C), service factor 1.15, FLA not greater than 9 amps, FLA 9.1 amps to 20 amps, FLA greater than 20 amperes, 1 hp or less (nonauto start), 1 hp or less (automatic start).

   A. Overload protection is:
      1. (FLA Nameplate) × 125 percent service factor 1.15, Table 430–32(1)(a)
      2. (Nameplate FLA) × 125 percent (40°C temperature rise)

   B. If the motor overload protection is not sufficient to start the motor or carry the load, it is permitted to be increased in accordance with Article 430–34.

   C. Thermal protection (internal overload protection) can be used in lieu of separate overloads in accordance with Article 430–7(a) 13, 430–32(a)(2).

Step 4

Branch circuit, short circuit and ground fault protection

Article 430–51, Part D, 240–3, 6

Overcurrent protection is based on Table 430–152.
The following are exceptions, Article 430–52(c)(1):

A. When calculated size overcurrent protection does not correspond to standard size, the next higher standard size or setting can be used.

B. When calculated size will not allow motor to start, the following can be applied.
1. Not greater than 600 volts or time delay class cc fuse can be increased by 400 percent of the FLA.
2. The rating of a dual element time delayed fuse can be increased by not greater than 225 percent of FLA.
3. Inverse time circuit breakers can be increased by 400 percent for protection 100 amps or less and 300 percent for protection greater than 100 amps.
4. FLA protection can be increased by 300 percent for fuses 601–6000 amperes.
5. Manufacturer's specifications for protection cannot be exceeded.
6. Instantaneous trip breakers can be used only if adjustable and part of a combination listed controller. In this case a multiplier of 1300 percent of FLA shall be allowed to start motor.
7. When using instantaneous trip circuit breakers an increase of (A) 1300 percent of FLA for other than design E motors, (B) no more than 1700 percent for design E motors, (C) 800 percent for other than design E motors under engineering evaluation, (D) 1100 percent for design E motors under engineering evaluation, can be used when calculations based on Table 430–152 do not allow the motor to start.
8. Listed combination motor controllers rated at less than 8 amperes with an instantaneous-trip circuit breaker and a continuous rating of 15 amperes can be increased by the value marked on the controller.
9. One overcurrent protection for the smallest winding can be used in multi-speed applications.
10. Torque motors can be protected with the name plate FLA instead of Table 430–150.

Step 5

Size feeder overcurrent protection
To size the feeder OCPD: Take the largest OCPD (fuse or breaker) plus (+) the sum of the full-load currents of all other motors.

Motor Control Circuits

- To size overcurrent protection for control transformers, see Table A–2 for primary side and Table A–3 for the secondary side.
- To size controllers for all applications *except* plugging and jogging, see Table A–4.
- To size controllers for plugging and jogging applications, see Table A–5.
- To determine wire size, see Table A–11.
- To size grounding conductors, see Table A–12.

TABLE A–1   Running Overcurrent Units[†]

Kind of Motor	Supply System	Number and Location of Overcurrent Units (Such as Trip Coils, Relays, or Thermal Cutouts)
1-phase AC or DC	2-wire, 1-phase AC or DC ungrounded	1 in either conductor
1-phase AC or DC	2-wire, 1-phase AC or DC, one conductor grounded	1 in ungrounded conductor
1-phase AC or DC	3-wire, 1-phase AC or DC, grounded-neutral	1 in either ungrounded conductor
3-phase AC	Any 3-phase	*3, one in each phase

*Exception: Unless protected by other approved means.
NOTE: For 2-phase power supply systems see the *National Electrical Code*®, Section 430–37.

TABLE A–2   Control Transformer Overcurrent Protection (Primary Voltage)[†]

Rated Primary (Amperes)	Maximum Rating of Primary Overcurrent Protective Device as a Percent of Transformer Rated Primary Current
Less than 2	300[1]
2 or more	250

[1]500% is permitted for a circuit of a control apparatus or system that carries the electric signals directing the performance of the motor controller, but does not carry the main power current.

TABLE A–3   Control Transformer Overcurrent Protection (120 Volt Secondary)[†]

Control Transformer Size, Volt-Amperes	Maximum Rating, Amperes
50	0.5
100	1.0
150	1.6
200	2.0
250	2.5
300	3.2
500	5
750	8
1000	10
1250	12
1500	15
2000	20
3000	30
5000	50

NOTE: For transformers larger than 5000 volt-amperes, the protective device rating shall be based on 125 percent of the secondary current rating of the transformer.

TABLE A–4   **Ratings for Three-Phase, Single-Speed, Full-Voltage Magnetic Controllers for Nonplugging and Nonjogging Duty**[†]

| Size of Controller | Continuous Current Rating[2] (Amperes) | Horsepower[1,4] at | | | | Service-Limit Current Rating[3] (Amperes) |
		60 Hertz 200 Volts	60 Hertz 230 Volts	50 Hertz 380 Volts	60 Hertz 460 or 575 Volts	
00	9	1½	1½	1½	2	11
0	18	3	3	5	5	21
1	27	7½	7½	10	10	32
2	45	10	15	25	25	52
3	90	25	30	50	50	104
4	135	40	50	75	100	156
5	270	75	100	150	200	311
6	540	150	200	300	400	621
7	810	—	300	—	600	932
8	1215	—	450	—	900	1400
9	2250	—	800	—	1600	2590

Reference: ANSI/NEMA ICS-2, Table 2-321-1.

[1]These horsepower ratings are based on locked-rotor current ratings given in ANSI/NEMA ICS-2, Table 2-237-3. For motors having higher locked-rotor currents, a larger controller should be used so that its locked-rotor current rating is not exceeded. (Refer to ANSI/NEMA ICS-2 for horsepower ratings of single-phase, reduced voltage, or multispeed motor controller application.)

[2]The continuous-current ratings represent the maximum rms current, in amperes. the controller may be expected to carry continuously without exceeding the temperature rises permitted by ANSI/NEMA ICS-1, Part 109 Part ICS 1-109.

[3]The service-limit current ratings represent the maximum rms current, in amperes, the controller may be expected to carry for protracted periods in normal service. At service-limit current ratings, temperature rises may exceed those obtained by testing the controller at its continuous current rating. The current rating of overload relays or the trip current of other motor protective devices used shall not exceed the service-limit current rating of the controller.

[4]Refer to ANSI/NEMA ICS-2 for horsepower ratings of single-phase, reduced voltage, or multispeed motor controller application.

TABLE A–5 **Ratings for Three-Phase, Single-Speed, Full-Voltage Magnetic Controllers for Plug-Stop, Plug-Reverse, or Jogging Duty**[†]

Size of Controller	Continuous Current Rating[2] (Amperes)	Horsepower[1,4] at				Service-Limit Current Rating[3] (Amperes)
		60 Hertz 200 Volts	230 Volts	50 Hertz 380 Volts	60 Hertz 460 or 575 Volts	
0	18	1½	1½	1½	2	21
1	27	3	3	5	5	32
2	45	7½	10	15	15	52
3	90	15	20	30	30	104
4	135	25	30	50	60	156
5	270	60	75	125	150	311
6	540	125	150	250	300	621

Reference: ANSI/NEMA ICS-2, Table 2-321-3.

[1]These horsepower ratings are based on locked-rotor current ratings given in ANSI/NEMA ICS-2, Table 2-237-3. For motors having higher locked-rotor currents, a larger controller should be used so that its locked-rotor current rating is not exceeded. (Refer to ANSI/NEMA ICS-2 for horsepower ratings of single-phase, reduced voltage, or multispeed motor controller application.)

[2]The continuous-current ratings represent the maximum rms current, in amperes, that the controller may be expected to carry continuously without exceeding the temperature rises permitted by ANSI/NEMA ICS-I, Part 109 Part ICS 1-109.

[3]The service-limit current ratings represent the maximum rms current, in amperes, the controller may be expected to carry for protracted periods in normal service. At service-limit current ratings, temperature rises may exceed those obtained by testing the controller at its continuous current rating. The current rating of overload relays or the trip current of other motor protective devices used shall not exceed the service-limit current rating of the controller.

[4]Refer to ANSI/NEMA ICS-2 for horsepower ratings of single-phase, reduced voltage, or multispeed motor controller application.

TABLE A–6   Color Coding for Push Buttons, Indicator (Pilot) Lights, and Illuminated Push Buttons†

Color	Device Type	Typical Function	Example
RED	Pushbutton	Emergency Stop, Stop, Off	Emergency Stop button, Master Stop button, Stop of one or more motors.
	Pilot Light	Danger or alarm, abnormal condition requiring immediate attention	Indication that a protective device has stopped the machine, e.g., overload.
	Illuminated Pushbutton		Machine stalled because of overload, etc. (The color RED for the emergency stop actuator shall not depend on the illumination of its light.)
YELLOW (Amber)	Pushbutton	Return, Emergency Return, Intervention—suppress abnormal conditions	Return of machine elements to safe position, override other functions previously selected. Avoid unwanted changes.
	Pilot Light	Attention, caution/marginal condition. Change or impending change of conditions	Automatic cycle or motors running; some value (pressure, temperature) is approaching its permissible limit. Ground fault indication. Overload that is permitted for a limited time.
	Illuminated Pushbutton	Attention or caution/Start of an operation intended to avoid dangerous conditions	Some value (pressure, temperature) is approaching its permissible limit; pressing button to override other functions previously selected.
GREEN	Pushbutton	Start-On	General or machine start; start of cycle or partial sequence; start of one or more motors; start of auxiliary sequence; energize control circuits.
	Pilot Light	Machine Ready; Safety	Indication of safe condition or authorization to proceed. Machine ready for operation with all conditions normal or cycle complete and machine ready to be restarted.
	Illuminated Pushbutton	Machine or Unit ready for operation/Start or On	Start or On after authorization by light; start of one or more motors for auxiliary functions; start or energization of machine elements.
BLACK	Pushbutton	No specific function assigned	Shall be permitted to be used for any function except for buttons with the sole function of Stop or Off; inching or jogging.
WHITE or CLEAR	Pushbutton	Any function not covered by the above	Control of auxiliary functions not directly related to the working cycles.
	Pilot Light	Normal Condition Confirmation	Normal pressure, temperature.
	Illuminated Pushbutton	Confirmation that a circuit has been energized or function or movement of the machine has been started/Start-On, or any preselection of a function	Energizing of auxiliary function or circuit not related to the working cycle; start or preselection of direction of feed motion or speeds.
BLUE or GRAY	Pushbutton, Pilot Light, or Illuminated Pushbutton	Any function not covered by the above colors	

For illuminated pushbuttons the function(s) of the light is separated from the function(s) of the button by a virgule (/).

TABLE A–7 Minimum Conductor Size[†]

Conductors shall not be smaller than:

Power circuits . . . . No. 14

Lighting and control circuits on the machine and in raceways . . . . No. 16

Exception: In a jacketed, multiconductor cable assembly, No. 18 shall be permitted.

Control circuits within control enclosures or operator stations . . . . No. 18

TABLE A–8 Single Conductor Characteristics[†]

Size (AWG/kcmil)	Cross-Sectional Area—Nominal (CM/mm²)	DC Resistance at 250° C (ohms/1000 ft)	Minimum Number of Strands		
			Nonflexing (ASTM Class)	Flexing (ASTM Class)	Constant Flex (ASTM Class/AWG Size)
22	640/0.324	17.2	7(¢)	7(¢)	19(M/34)
20	1020/0.519	10.7	10(K)	10(K)	26(M/34)
18	1620/0.823	6.77	16(K)	16(K)	41(M/34)
16	2580/1.31	4.26	19(C)	26(K)	65(M/34)
14	4110/2.08	2.68	19(C)	41(K)	41(K/30)
12	6530/3.31	1.68	19(C)	65(K)	65(K/30)
10	10380/5.261	1.060	19(C)	104(K)	104(K/30)
8	16510/8.367	0.6663	19(C)	(\)	(-)
6	26240/13.30	0.4192	19(C)	(\)	(-)
4	41740/21.15	0.2636	19(C)	(\)	(-)
3	52620/26.67	0.2091	19(C)	(\)	(-)
2	66360/33.62	0.1659	19(C)	(\)	(-)
1	83690/42.41	0.1315	19(B)	(\)	(-)
1/0	105600/53.49	0.1042	19(B)	(\)	(-)
2/0	133100/67.43	0.08267	19(B)	(\)	(-)
3/0	167800/85.01	0.06658	19(B)	(\)	(-)
4/0	211600/107.2	0.05200	19(B)	(\)	(-)
250 kcmil	-/127	0.04401	37(B)	(\)	(-)
300	-/152	0.03667	37(B)	(\)	(-)
350	-/177	0.03144	37(B)	(\)	(-)
400	-/203	0.02751	37(B)	(\)	(-)
450	-/228	0.02445	37(B)	(\)	(-)
500	-/253	0.02200	37(B)	(\)	(-)
550	-/279	0.02000	61(B)	(\)	(-)
600	-/304	0.01834	61(B)	(\)	(-)
650	-/329	0.01692	61(B)	(\)	(-)
700	-/355	0.01572	61(B)	(\)	(-)
750	-/380	0.01467	61(B)	(\)	(-)
800	-/405	0.01375	61(B)	(\)	(-)
900	-/456	0.01222	61(B)	(\)	(-)
1000	-/507	0.01101	61(B)	(\)	(-)

(B, C, K) ASTM Class designation B and C per ASTM B8-86, Class designation K per ASTM B 174–71 (R1980).

(') A class designation has not been assigned to this conductor but is designated as size 22-7 in ASTM B286-1989 (R1979) and is composed of strands 10 mils in diameter (No. 30 AWG).

(\) Nonflexing construction shall be permitted for flexing service * Per ASTM Class designation B 174–71 (R1980), Table 3.

(-) Constant flexing cables are not constructed in these sizes.

TABLE A–9 **Single Conductor Insulation Thickness of Insulation in Mils***
[Average/Minimum (Jacket)]†

Wire Size	A	B
22 AWG	30/27	15/13(4)
20	30/27	15/13(4)
18	30/27	15/13(4)
16	30/27	15/13(4)
14	30/27	15/13(4)
12	30/27	15/13(4)
10	30/27	20/18(4)
8	45/40	30/27(4)
6	60/54	30/27(5)
4–2	60/54	40/36(6)
1–4/0	80/72	50/45(7)
250–500 MCM	95/86	60/54(8)
550–1000	110/99	70/63(9)

*UL 1063 Table 1.1 *NEC®* Construction A-No outer covering B-Nylon covering.

TABLE A–10   **Conductor Ampacity Based on Copper Conductors with 60° C and 75° C Insulation in an Ambient Temperature of 30° C†**

Conductor Size AWG	Ampacity in Cable or Raceway		Control Enclosure 60° C*
	60° C	75° C	
30	—	0.5	0.5
28	—	0.8	0.8
26	—	1	1
24	2	2	2
22	3	3	3
20	5	5	5
18	7	7	7
16	10	10	10
14	15	15	20
12	20	20	25
10	30	30	40
8	40	50	60
6	55	65	80
4	70	85	105
3	85	100	120
2	95	115	140
1	110	130	165
0	125	150	195
2/0	145	175	225
3/0	165	200	260
4/0	195	230	300
250	215	255	340
300	240	285	375
350	260	310	420
400	280	335	455
500	320	380	515
600	355	420	575
700	385	460	630
750	400	475	655
800	410	490	680
900	435	520	730
1000	455	545	780

*Sizing of conductor in wiring harnesses or wiring channels shall be based on the ampacity for cables.

NOTE 1: Wire types listed in 15.1 shall be permitted to be used at the ampacities as listed in this table.

NOTE 2: For ambient temperatures other than 30° C, see *NEC*® Table 310–16 correction factors.

NOTE 3: The sources for the ampacities in this table are Table 310–16 and 310–17 of the *NEC*®.

TABLE A–11   Maximum Conductor Size for Given
Motor Controller Size[†]

Motor Controller Size	Maximum Conductor Size AWG or MCM
00	14
0	10
1	8
2	4
3	0
4	000
5	500

*See ANSI/NEMA ICS 2-1988 Table 2, 110–1.

TABLE A–12   Size of Grounding Conductors[†]

Column "A," Amperes	Copper Conductor Size, AWG
10	16* or 18*
15	14, 16*, or 18*
20	12, 14*, 16*, or 18*
30	10
40	10
60	10
100	8
200	6
300	4
400	3
500	2
600	1
800	0
1000	2/0
1200	3/0
1600	4/0

*Permitted only in multiconductor cable where connected to portable
or pendant equipment.

APPENDIX

B

Communication and Control Station Identification and Operation

Color designation is an important factor in push-button switch operators. Color designation not only provides an attractive panel but, more important, it lends itself to safety.

Quick identification is important. Certain functions soon become associated with a specific color. Standards have been developed that specify certain colors for particular functions. For example, in the machine tool industry, the colors red, yellow, and black are assigned the following functions:

Red Stop, Emergency Stop
Yellow Return, Emergency Return
Black Start Motors, Cycle

A control station may contain push buttons alone or a combination of push buttons, selector switches, and pilot lights (indicating lights). Indicating lights may be mounted in the enclosure. These lights are usually red or green and are used for communication and safety purposes. Other common colors available are amber, blue, white, and clear. They indicate when the line is energized, the motor is running, or any other condition is designated.

The indicating or pilot light is available in three basic types:

1. Full voltage.

2. Resistor.

3. Transformer.

Due to the vibration normally present in machines, the low-voltage bulb is preferred. The low-voltage bulb operates at 6 volt to 8 volt obtained through a resistance or transformer unit.

The lens is available either in plastic or glass and in a variety of colors. Again, as for push-button operators, colors can be used for selection purposes to increase safety of operation. For example:

Red Danger, Abnormal Conditions
Amber Attention

Green   Safe Condition

White or Clear   Normal Condition

The push-to-test indicating light provides an additional feature. Consider, for example, an indicating lamp that will not illuminate. It may be that the bulb is not energized, or the bulb may be burned out. Depressing the lens unit will connect the bulb directly across the control voltage source. This provides a check on the condition of the bulb.

Note that the push-to-test pilot light consists of a 120-6 volt transformer, a 6-volt bulb covered by a lens, and a standard contact block. When the pilot light lens is depressed (operated), it mechanically operates the contact block. The normally closed contacts open and the normally open contacts close. The normally closed contacts are connected to the control circuit. The normally open contacts are connected directly to a source of 120 volts. Thus, when the lens is depressed (operated) the pilot light transformer primary is connected directly across 120 volts. Six volts is now applied to the bulb.

Miniature Oil-Tight Units

In addition to the standard line of indicating lights, a line of miniature oil-tight push buttons, selector switches, and pilot lights is now available. This line covers about the same selection as the standard line. These units can often be used to an advantage where space is at a premium. This is particularly true where a great deal of indicating is required.

Control stations may also include switches that are key, coin, or hand operated wobble sticks. A wobble stick is a stem-operated push button, for operation from any direction. There are ball lever push button operators for a gloved hand or for frequent operations.

Name plates are installed to designate each control operation. These control operators are commonly used in control circuits of magnetic devices in factory production machinery.

Combination indicating light, nameplate, push-button units are available. These illuminated push buttons and indicating lights are designed to save space in a wide variety of applications such as control and instrument panels, laboratory instruments, and computers. Miniature buttons are also used for this purpose.

Standard control station enclosures are available for normal general-purpose conditions, while special enclosures are used in situations requiring water-tight, dust-tight, oil-tight, explosion-proof, or submersible protection. Provisions are often made for padlocking stop buttons in the open position (for safety purposes). Relays, contactors, and starters cannot be energized while an electrician is working on them with the stop button in this position.

APPENDIX

C

Resources for Motor Control Design and Applications

The following are contacts that can provide you with information about their products and help you with design and applications beyond the scope of this book.

Drives

Allen Bradley
Siemens
Square D

PLCs

Allen Bradley
Siemens
Square D

Distributive Controls

Allen Bradley
Siemens
Square D

Troubleshooting, Circuitry, and Consulting

Integrity Electrical Services
Siemens
Square D

Industry Standards

Siemens
Square D
Department of Energy—Motor Challenge
NFPA 70 & 79

Motors

Siemens
Department of Energy—Motor Challenge
NFPA

Software

Power Quality
Integrity Software

Motor Calculation

Integrity Software

Names and Addresses

Allen Bradley
Box 760
Mequon, WI 53092
Tel: 414-242-8200 Fax: 414-242-8665

Department of Energy—Motor Challenge
Washington State Energy Office
Box 43165
Olympia, WA . 98504-3165
Tel: 360-956-2034
Fax: 360-956-2229

Integrity Software
60 Evergreen Park
Florence, AL 35633
Email: integrityco@mindspring.com
Tel: 256-718-3320

Integrity Electrical Services
60 Evergreen Park
Florence, AL 35633
Tel: 256-718-3320

NFPA
Batterymarch Park
Quincy, MA 02269
Tel: 617-770-3000
Fax: 617-770-0700

Siemens
Siemens Energy and Automation
100 Technical Drive
Alpharetta, GA 30202

Square D Company
Automation and Control Business
P.O. Box 27446
Raleigh, NC 27611

APPENDIX

D

Common Control Schematics

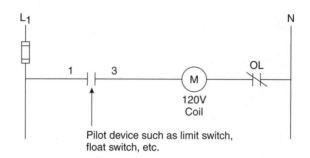

Pilot device such as limit switch,
float switch, etc.

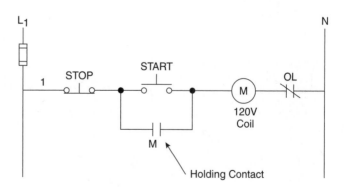

Holding Contact

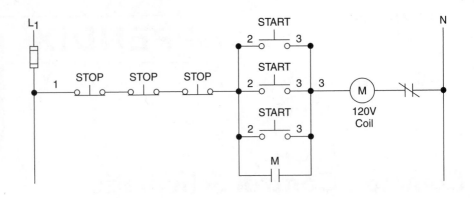

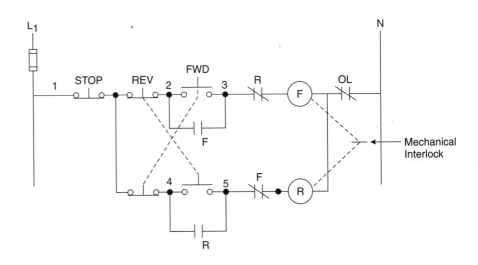

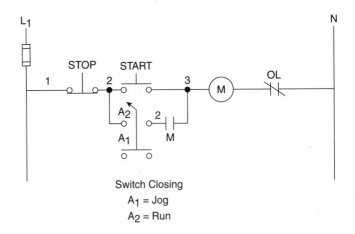

Switch Closing
A_1 = Jog
A_2 = Run

APPENDIX

Symbols

Switches				
Disconnect	Circuit Interrupter	Circuit Breaker W/Thermal O.L.	Circuit Breaker W/Magnetic O.L.	Circuit Breaker W/Thermal and Magnetic O.L.

Limit Switches		Foot Switches	Pressure and Vacuum Switches		Liquid Level Switches	
Normally Open	Normally Closed	NO	NC	NO	NC	NO

Held Closed	Held Open	NC	Temperature Actuated Switches		Flow Switches (Air, Water, Etc.)	

Speed (Plugging)		Anti-Plug	Selector			
F	F	F	2 Position	3 Position	2 Pos. Sel. Pushbutton	

2 Position

J — K

○A1 | J | K |
 | A1 | X | |
○A2 | A2 | | X |

X - Contact Closed

3 Position

J K L

○A1 | J | K | L |
 | A1 | X | | |
○A2 | A2 | | | X |

X - Contact Closed

2 Pos. Sel. Pushbutton

A — B

1○ ○2
3○ ○4

Contacts	Selector Position	
	A	B
	Button	Button
	Free	Depres'd	Free	Depres'd
1-2	X			
3-4		X	X	X

Pushbuttons							
Momentary Contact					Maintained Contact		Illuminated
Single Circuit		Double Circuit	Mushroom Head	Wobble Stick	Two Single Circuit	One Double Circuit	
NO	NC	NO & NC					(R)

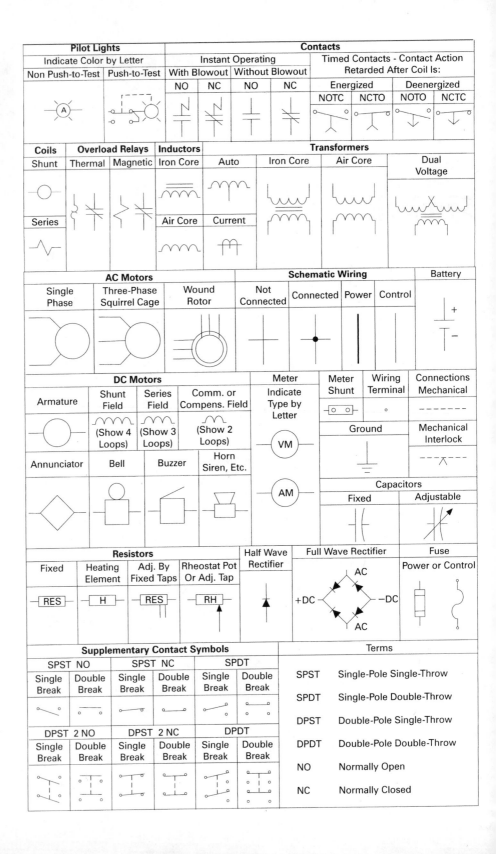

Testing an SCR

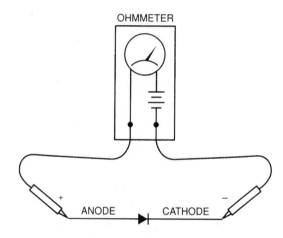

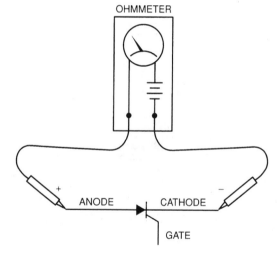

(1) Using a junction diode, determine which ohmmeter lead is positive and which is negative. The ohmmeter will indicate continuity only when the positive lead is connected to the anode of the diode and the negative lead is connected to the cathode.

(2) Connect the positive ohmmeter lead to the anode of the SCR and the negative lead to the cathode. The ohmmeter should indicate no continuity.

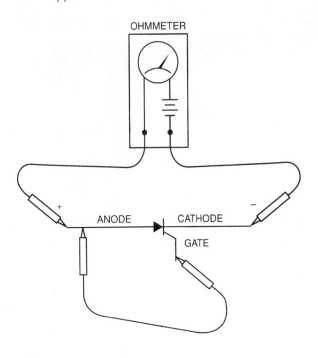

(3) Using a jumper lead, connect the gate of the SCR to the anode. The ohmmeter should indicate a forward diode junction when the connection is made. NOTE: If the jumper is removed, the SCR may continue to conduct or it may turn off. This will be determined by whether or not the ohmmeter can supply enough current to keep the SCR above its holding current level.

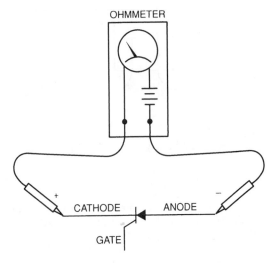

(4) Reconnect the SCR so that the cathode is connected to the positive ohmmeter lead and the anode is connected to the negative lead. The ohmmeter should indicate no continuity.

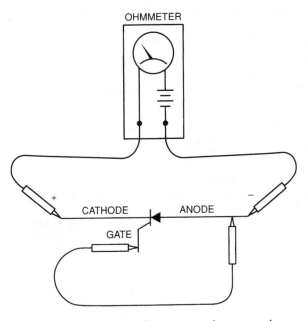

(5) If a jumper lead is used to connect the gate to the anode, the ohmmeter should indicate no continuity. NOTE: SCRs designed to switch large currents (50 amperes or more) may indicate some leakage current with this test. This is normal for some devices.

APPENDIX

G

Common Troubleshooting Points

It would be impractical, if not impossible, to list all potential trouble spots. However, the following are common areas that contribute to a large percentage of troubles. Check the following before evaluating other problems. *Refer to the flowchart at the end of this appendix while reviewing the material.*

Fuses and Circuit Breakers

The checking of fuses is generally a good place to start. Too often this is overlooked. Fuses blow for one or more of the following reasons:

1. Overloads.
2. Short-circuits.
3. Ground-faults.

NOTE: Under no circumstances should you replace the fuse or circuit breaker without having a reasonable answer as to why the fuse blew.

The replacement of an open (defective) fuse can be an important safety factor. There are three different types of fuses. Within each type are different voltage and current ratings. Too often just any type of fuse is used as a replacement. Unless changes have been made in a machine circuit and components, the replacement fuse should be exactly the same type, voltage, and current rating as the fuse removed.

The policing of fuses can be a problem. However, the replacement policy given here must be rigidly maintained if safety to personnel and the equipment is to be realized.

One extremely important case involves a machine connected to a power source that has a high short-circuit current available. In such cases, it may be advisable to use current-limiting fuses with high interrupting capacity.

Loose Connections

Each connection may be a source of trouble. Much advancement has been made in terminal block and component connectors to improve this condition. The use of stranded conductors in place of solid conductors has in general improved the connection problem.

The problem starts when the equipment is installed and continues throughout the life of the machine. It may be of greater importance in power circuits, as the current handled is of greater magnitude. A loose correction in a power circuit can generate local heat. This spreads to other parts of the same component, other components, or conductors. An example of where direct trouble can arise is thermally sensitive elements. These can be overload relays or thermally operated circuit breakers.

For the correction of loose connections, the best advice is to follow a good program of preventive maintenance in which connections are periodically checked and tightened. Check with the manufacturer's specifications for torque requirements or see *Electrician's Technical Reference: Motors* for proper torquing techniques and specifications.

Faulty Contacts

This applies to such components as motor starters, contactors, relays, push buttons, and switches.

A problem that appears quite often and one of the most difficult to locate is the normally closed contact. Observation indicates that the contact is closed but does not reveal if it is conducting current.

Any contact that has had an overload through it should be checked for welding.

Such conditions as weak contact pressure and dirt or an oxide film on contact will prevent it from conducting. Contacts may be cleaned by drawing a piece of rough paper between the contacts.

> **CAUTION:** Use only a fine abrasive to clean contacts. Do not file contacts. Most contacts have a silver plate over the copper. If this is destroyed by filing, the contact will have a short life. If contacts are worn or pitted so badly that a fine abrasive will not clean them, it is better to change the contacts.

Another problem that may occur with a double-pole, double-break contact is crossfiring. That is, one contact of the double break travels across to the opposite contact, but the other remains in its original position. If both the NO and NC contacts are being used in the circuit, a malfunction of control may occur.

Incorrect Wiremarkers

This problem usually occurs during installation, manufacturing, remolding, or in reassembly in the user's plant. The error can be difficult to locate, as a cable may have many conductors running some distance to various parts of the equipment.

One common problem is the transposition of numbers. For example, a conductor may have a 69 marked on one end and a 96 on the other end. Another problem that may occur is in connecting conductors into a terminal block. With a long block and many conductors, it is a common error to connect a conductor either one block above or below the proper position.

Combination Problems

Reference has been made to combination problems, but their importance should be emphasized. The following are typical types of combination problems:

- Electrical-mechanical
- Electrical-pressure (fluid power or pneumatic)
- Electrical-temperature

The greatest problem is that the observed or reported trouble is not always indicative of which aspect is at fault. It may be both.

It is usually faster to check the electrical circuit first. However, both systems must be checked as both may contribute to the problem.

As an example, very few solenoid coils burn out due to a defect in the coil. Probably over 90 percent of all solenoid trouble on valves develops from a faulty mechanical or pressure condition that prevents the solenoid plunger from seating properly, thus drawing excessive current. The result is an overload or a burned-out solenoid coil.

Low Voltage

If no immediate indication of trouble is apparent, one of the first checks to make is the line and control voltage. Due to inadequate power supply or conductor size, low voltage can be a problem.

The problem generally shows up more on starting or energizing a component, such as a motor starter or solenoid. However, it can cause trouble at other spots in the cycle.

A common practice in small shops is to add more machines without properly checking the power supply (line transformers) or the line conductors. The source and line become so heavily loaded that when they are called on for a normal temporary machine overload, the voltage drops off rapidly. This may result in magnetic devices such as starters and relays dropping off the line (opening their contacts) through undervoltage or overload protective devices.

Heat is one result of low voltage that may not be noticed immediately in the functioning of a machine. As the voltage drops, the current to a given load increases. This produces heat in the coils of the components (motor starters, relays, solenoids) which not only shortens the life of the components but may cause malfunctioning. For example, where there are relatively moving metal parts with close tolerances, heat can cause these parts to expand to a point of sticking. In cases in which electrical heating is used, the heat is reduced by the square of the voltage. For example, if the voltage is dropped to one half of the heating element's rated voltage, the heat output will be reduced to one fourth.

Grounds

Typical Locations

There are many locations on a machine where a grounded condition can occur. However, there are a few spots in which grounds occur most often.

- *Connection points in solenoid valves, limit switches, and pressure switches.* Due to the design of many components, the space allowed for conductor entrance and connection is limited. As a result, a part of a bare conductor may be against the side of an uninsulated component case. Where bolt and nut connections are made, the insulating tape may not be wrapped securely, or it may be of such quality that age destroys its insulating properties. This may occur in the field (user's plant), where due to the urgency for a quick change, sufficient care is not taken when handling the wiring and connections on a replaced unit.

- *Pulling conductors.* In pulling conductors through conduit where there are several bends and 90° fittings, or into pull boxes and cabinets, the conductor insulation may be scraped or cut. If care is not taken to eliminate sharp edges or burrs on freshly cut or machined parts, cuts and abrasions occur. This is one good reason for the use of the insulation required for machine tool wiring.

- *Loose strands.* The use of stranded wire has greatly reduced many problems in machine wiring. However, care must be taken when placing a stranded conductor into a connector. All strands must be used. One or two strands unconnected can touch the case or a normally grounded conductor, creating an unwanted ground. Even if the ground condition does not appear, the current-carrying capacity of the conductor is reduced.

Means of Detection

For the best operation of an electrical system, some means of detecting the presence of grounds should be available. There are two methods detailed here. Each has merit.

In the first method, all coils are tied solidly to a common line, and this line is grounded. The opposite side of the control power source is protected by a fuse or circuit breaker. When the ground is located and removed, the fuse can be replaced or the circuit breaker reset. Thus, the control circuit is again ready for operation. Note that the ground condition must be removed first.

In the second method, measure continuity from lead to ground.

Wiring and Grounding

Incorrect wiring and grounding can result in major problems for companies with computers and computer controlled equipment. These problems often show up as unexpected shut downs, burned out equipment, scrambled data, and equipment lockups.

The Electric Power Research Institute estimates that 80 percent of all power quality is the result of poor grounding, wiring, or load placement on the user's circuits. By making a few simple checks with your DMM you may be able to eliminate some common power quality problems caused by wiring and grounding.

It is recommended that all circuits that serve critical loads should be on dedicated circuits and neutrals. Check wiring for correct bonding, tied neutrals and grounds, overloaded neutrals, shared neutrals to critical loads, and insuring that the neutral is not used for equipment grounding.

This may be all that is necessary to eliminate many problems associated with poor equipment performance.

Poor Housekeeping

Poor "housekeeping" leads to more work for the troubleshooter. There is an overall economy in having a clean machine and a well-organized and well-executed preventive maintenance program.

Dust, dirt, and grease should be removed periodically from electrical parts. Their presence only causes mechanical failure and forms paths between points of different potential, causing a short circuit.

Moving mechanical parts should be checked, particularly in large motor starters. Such items as loose pins and bolts and wearing parts are sources of trouble.

Overheated parts generally indicate trouble. Without proper instruments, it is difficult to determine the temperature of a part or how high a temperature can be sustained. Certainly any signs of smoke or baking of insulation are cause for immediate concern.

Manufacturers of components have done considerable design work to prevent dust, dirt, and fluids from entering.

When it is necessary to remove a cover or open a door for troubleshooting, immediately replace it after the trouble is corrected.

Many users have gone to great lengths to develop and rigidly enforce a good electrical maintenance program. Records are kept of each reported trouble and the work is done to correct the problem. These records are compiled periodically and are available to the supervisor. This not only leads to faster troubleshooting in the future, but it also gives the supervisor an indication of why the production in a given department may be down.

Trouble Patterns

As troubleshooting work progresses over a period of time with a particular machine or group of machines, a pattern of trouble may develop. For example, it may be necessary to increase the production rate with a particular machine. Certain areas of this machine may not have been originally designed to handle this increased rate of operation. Unless the machine is redesigned and rebuilt, a pattern of trouble may develop.

Another example is a machine that is relocated in another section of the user's plant. Here the environment may be different. A change in the atmosphere and the presence of dust, dirt, or metal chips may create a pattern of trouble for a machine if it is not designed to operate under these conditions.

Opens in a Common Line

Many circuits have two or more common connection points for multiple connections.

In a rather simple circuit, it does not appear that this error would be committed. However, there are cases in which there may be many of these common connections. In larger and more complicated circuit diagrams, the probability of jumpers being missed is much greater.

Wiring the Wrong Contact

Many components such as relays, limit switches, and pressure and temperature switches have NO and NC contacts available for use.

A wiring error is often made particularly when only one of the two available contacts is used. The error consists of wiring the wrong side of the contact; that is, the NO contact may be put into the circuit where the NC should be used. The reverse of this may also be true.

Unless the person who does the wiring is completely familiar with the component and double-checks the work, this error will occur frequently.

The routine checks discussed in the following section will reveal this error quickly.

Momentary Faults

The *momentary fault* is one of the most difficult problems to troubleshoot. With this type of fault, the machine or control can be under close observation for hours with no failures. The fault may occur at any time, however, and if the observer's attention is even briefly diverted, any direct evidence that might have been seen is lost.

There is no direct solution to this problem. The best approach is a well-organized analysis. The following steps might be helpful:

1. Attempt to localize within the total cycle. If the fault always occurs at the same place in the cycle, generally only the control associated with that part of the cycle is involved. If the fault occurs at random spots through the cycle, then the spots to examine are those that are common through the entire cycle. Examples are a drum or selector switch used to isolate an entire section of control.

2. Examine for loose connections, particularly in the area of the fault. Attention should be paid to areas where mechanical action may have damaged conductors, pulled them loose from connectors, or cut or broken them. The complete break of a stranded conductor within the insulation is rare.

3. Localize attention to components. Sometimes casual observation will not disclose the trouble. In these cases the complete replacement of the components in question is the quickest and best solution. Here, the plugging components have a distinct advantage in returning a machine to operating condition in a minimum of time.

4. Examine the circuit for unusual conditions. This type of trouble rarely occurs in the user's plant. However, there are cases in which, either through an oversight on the part of the circuit designer or by a change of operating conditions on the machine, a circuit change is indicated as a solution to the problem.

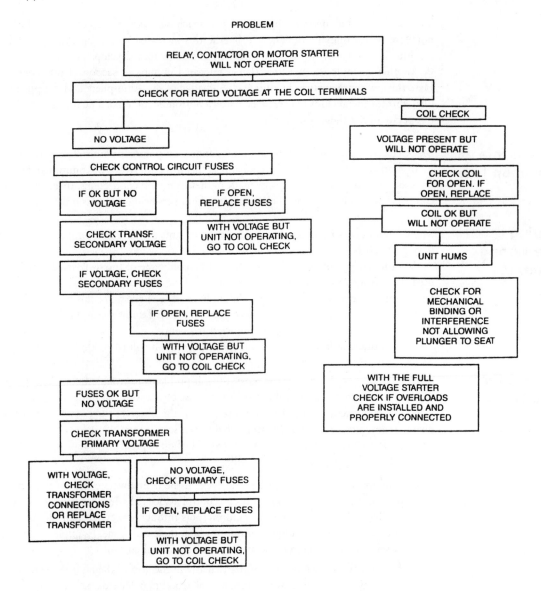

APPENDIX

NFPA Tables

TABLE H-1 One Line Representation of Electrical System Power Distribution*

Line	Reference	Single motor (main SCPD)	Multiple motors (main and branch SCPD)				Multiple motors (main SCPD only)
A	Supply NEC, Article 670						
B	Disconnecting means; Clause 7	Disc					Disc
C	Main overcurrent protection (when supplied) 8.2						
D	Branch overcurrent protection (when supplied) 8.2 or 8.3	See Figure 2	See Figure 2				See Figure 2
E	Control circuit and special purpose control protection cond: 8.9 trans. 8.12 undervoltage 8.14						
F	Motor control Clause 10	M	M	M	M	M	M
G	Overload protection 8.6						
	Disconnect (if used)	Disc	Disc	Disc	Disc	Disc	Disc
H	Motor and resistive loads Clause 16	MTR	MTR	MTR	MTR	MTR	MTR
		Each motor over 5 hp	Two motors 1 hp to 5 hp	Several motors under 1 hp	Several motors under 1 hp	Several motors under 1 hp	Several motors under 1 hp

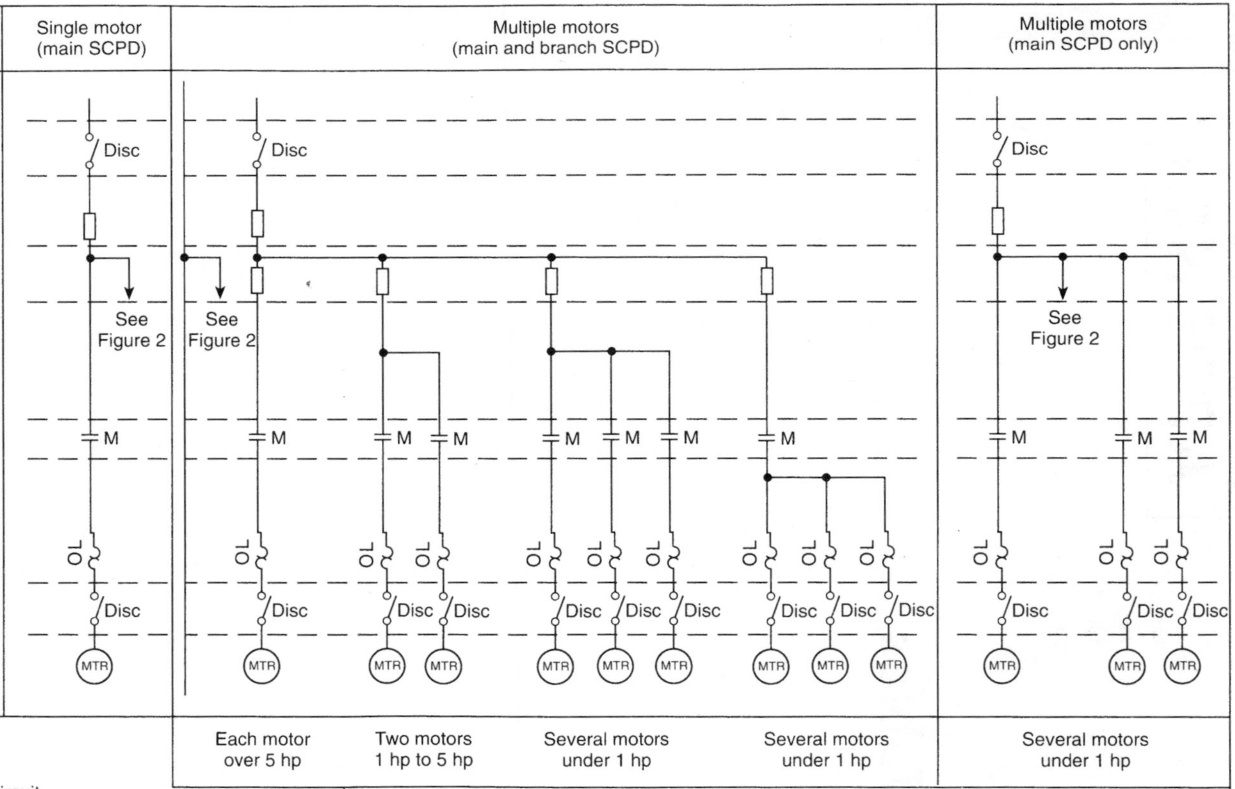

These applications may not provide type-2 protection for all devices.

Symbol ⊡ Short circuit protection device (SCPD)

TABLE H-2 One Line Representation of Electrical System Power Protection*

Line	Reference	Control circuit connections	Misc. drive connections
A	Supply NEC 70, Article 670		
B	Disconnecting means Clause 7		
C	Main overcurrent protection (when supplied) 8.2		
D	Overcurrent protection 8.2 or 8.3		
E1	Control circuit and special control protection cond., 8.9 trans., 8.12 undervoltage, 8.14		
E2	Externally protected drive unit		
E3	Internally protected drive unit		
F	Remote location drive unit internally or externally protected		
H	Motor Clause 16		

TABLE H–3   **Fuse and Circuit Breaker Selection: Motor, Motor Branch Circuit, and Motor Controller*[*]**

	Maximum Setting or Rating[1] (Fuse and Circuit Breaker)		
Application	Type[2]		
Fuse class with time delay[3]	AC-2	AC-3	AC-4
RK-5[4]	150	175	175
RK-1	150	175	175
J	150	175	225
CC	150	300	300
Instantaneous trip C/B[5]	700	700	700
Inverse trip C/B[6]	150	250	250

NOTES:

[1]For motors with locked-rotor code letters a-e, see ANSI/NFPA 70, Table 430–152.

[2]Types:

 AC-2: Slip-ring motors starting, switching off, or all light-starting duty motors.

 AC-3: Squirrel-cage motors; starting, switching off while running, occasional inching, jogging, or plugging but not to exceed 5 operations per minute or 10 operations per 10 minutes. All wye-delta and two-step auto-transformer starting, or all medium starting duty motors.

 AC-4: Squirrel-cage motors; starting, plugging, inching, jogging, or all heavy starting duty motors.

[3]Where the rating of a time delay fuse (other than CC type) specified by the table is not sufficient for the starting of the motor, it shall be permitted to be increased but shall in no case be permitted to exceed 225 percent.

[4]Class RK-5 fuses shall be used only with NEMA rated motor controllers.

[5]Magnetic only circuit breakers are limited to single motor applications. These instantaneous trip circuit breakers shall only be used if they are adjustable, if part of a combination controller having motor-running and also short-circuit and ground-fault protection in each conductor, and if the combination is especially identified for use, and it is installed per any instructions included in its listing or labeling. Circuit breakers with adjustable trip settings shall be set at the controller manufacturer's recommendation, but not greater than 1300 percent of the motor full-load current.

[6]Where the rating of an inverse time circuit breaker specified in the table is not sufficient for the starting current of the motor, it shall be permitted to be increased but in no case exceed short-circuit protective device (SCPD) as specified by the controller manufacturer. The maximum rating of the designated SCPD shall be as shown in the table.

*Reprinted with permission of NFPA 79, *Electrical Standard for Industrial Machinery,* National Fire Protection Association, Quincy, MA 02269. This reprinted material is not the official position of the National Fire Protection Association, which is represented by the standard in its entirety.

REFERENCES

The following documents are referenced for informational purposes only. The edition indicated for each reference is the current edition as of the date of NFPA issuance. These references will be helpful to the design, troubleshooting, and installation of motor controls. (ANSI Publications: American National Standards Institution, 11 W. 42nd Street, New York, NY 10036.)

ANSI B 11.19–1990, *Performance Criteria for the Design, Construction, Care, and Operation of Safeguarding when Referenced by Other B11 Machine Tool Safety Standards.*

ANSI B 11.20–1992, *Safety Requirements for the Construction, Care, and Use of Machine Tools—Manufacturing Systems and Cells.*

ANSI/NFPA 70E–1988, *Standard for Electrical Safety Requirements for Employee Workplaces.*

EIA RS-281-B (1979), *Electrical Construction Standards for Numerical Machine Control.*

ANSI/EIA RS-431 (1992), *Electrical Interface between Numerical Controls and Machine Tools.*

ANSI/IEEE STD 914984, *Graphical Symbols for Logic Functions.*

ANSIJEEE STD 100–1984, *Standard Dictionary of Electrical and Electronics Terms.*

IEC 204-1-1982 (2nd Edition), *Electrical Equipment of Industrial Machines.*

IEC 417 G-1985, *Graphical Symbols for Use on Equipment, 7th Supplement.*

IEC 550–1977, *Interface between Numerical Control and Industrial Machines.*

ANSI/NEMA ICS-1-1988, *Standards for Industrial Control and Systems.*

ANSI/NEMA ICS-2-1988, *Standards for Industrial Control Devices, Controllers, and Assemblies.*

NEMA MG-1-1978, *Motor and Generator Standards.*

ANSI C84. 1-1989, *Voltage Ratings for Electrical Power Systems and Equipment (60 Hz).*

ANSI/IEEE 315-1989, *Graphical Symbols for Electrical and Electronics Diagrams.*

ANSI/NEMA ICS-2: 1988, *Industrial Control Devices, Controllers, and Assemblies.*

ANSI/NFPA 70–1993, *National Electrical Code.*

ANSI/UL 1063–1986, *Standard for Safety Machine—Tool Wires and Cables.*

ASTM B 286–1989, *Copper Conductors for Use in Hookup Wire for Electronic Equipment.*

IEC 2041: 1992, *Electrical Equipment of Industrial Machines—Part 1: General Requirements.*

IEC 417G: 1985, *Graphical Symbols for Use on Equipment—Index, Survey, and Single Sheets.*

IEV 441: 1984, *International Electrotechnical Vocabulary—Chapter 441: Switchgear, Control Gear, and Fuses.*

IEV 826: 1982, *International Electrotechnical Vocabulary—Chapter 826: Electrical Installation of Buildings ISO 3864: 1984, Safety Colors and Safety Signs.*

GLOSSARY

AC Input Interface (Module): An input circuit that conditions various AC signals from connected devices to logic levels required by the processor.

AC Output Interface (Module): An output circuit that switches the user-supplied control voltage required to control connected AC devices.

Accessible (as applied to equipment): Admitting close approach because not guarded by locked doors, elevation, or other effective means.

Across-the-Line Starter: A motor starter that connects the motor to fun voltage supply.

Actuator: A cam, arm, or similar mechanical or magnetic device used to trip limit switches.

Address: A reference number assigned to a unique memory location. Each memory location has an address, and each address has a memory location.

Alphanumeric: A character set that contains both numerical and alphabetic characters.

Alternating Current (AC): Electric current that periodically changes direction and magnitude.

Alternation: One half-cycle in alternating current, either positive or negative half.

Ambient Conditions: The condition of the atmosphere adjacent to the electrical apparatus; the specific reference may apply to temperature, contamination, humidity, etc.

Ampacity: The current-carrying capacity, expressed in amperes.

Ampere (A): The unit of current.

Analog Device: Apparatus that measures continuous information (e.g., current voltage). The measured analog signal has an infinite number of possible values. The only limitation on resolution is the accuracy of the measuring device.

Analog Input Interface: An input circuit that employs an analog-to-digital converter to convert an analog value, measured by an analog measuring device to a digital value that can be used by the processor.

Analog Output Interface: An output circuit that employs a digital-to-analog converter to convert a digital value sent from the processor to an analog value that will control a connected analog device.

Analog Signal: One having the characteristic of being continuous and changing smoothly over a given range, rather than switching suddenly between certain levels, as with discrete signals.

Analog-to-Digital Conversion: Hardware and/or software process that converts a scaled analog signal or quantity into a scaled digital signal or quantity.

AND: An operation that yields a logic "I" output if all the inputs are "I" and a logic "0" if any of the inputs are "0".

Apparatus: The set of control devices used to accomplish the intended control functions.

Arithmetic Capability: The ability to perform such math functions as addition, subtraction, multiplication, division, square roots, etc. A given controller may have some or all of these functions.

Auxiliary Contacts: Contacts in addition to the main circuit contacts in a switching device. They function with the movement of the main circuit contacts.

Auxiliary Device: Any electrical device other than motors and motor starters necessary to fully operate the machine or equipment.

Available Short-Circuit Current: The maximum short-circuit current that can flow in an unprotected circuit.

Block Diagram: A diagram that shows the relationship of separate subunits (blocks) in the control system.

Bonding: The permanent joining of metallic parts to form an electrical conductive path that will assure electrical continuity and the capacity to conduct safely any current likely to be imposed (NFPA 79-1984, National Electrical Code).

Boolean Algebra: A mathematical shorthand notation that expresses logic functions, including AND, OR, EXCLUSIVE-OR, NAND, NOR, and NOT.

Branch Circuit: That portion of a wiring system extending beyond the final overcurrent device protecting the circuit. (A device not approved for branch circuit protection, such as a thermal cutout or motor overload protective device, is not considered the overcurrent device protecting the circuit).

Breakdown Voltage: The voltage at which a disruptive discharge takes place, either through or over the surface of insulation.

Breaker: An abbreviated name for the circuit breaker.

Bus: Power distribution conductors.

CMOS: An abbreviation of *complimentary metal oxide semiconductor.* A family of very low power, high-speed integrated circuits.

Capacitor: Two conductors separated by an insulator.

Coaxial Cable: A cable that is constructed with an outer conductor forming a cylinder around a central conductor. An insulating dielectric separates the inner and outer conductors. The complete assembly is enclosed in a protective outer sheath.

Combination Starter: A magnetic motor starter with a manually operated disconnecting device built into the same enclosure with the motor starter. Protection is always added. A control transformer may be added to provide 120 volts for control. START-STOP pushbuttons and a pilot light may be added in the enclosure door.

Command: In data communication, an instruction represented in the control field or a frame and transmitted by the primary device. It causes the addressed secondary to execute a specific data link control function.

Communication Network: A communication system that consists of adapter modules and cabling. It is used to transfer system status and data between network programmable logic controllers and computers.

Compartment: A space within the base, frame, or column of the equipment.

Compatibility: The ability of various specified units to replace one another with little or no reduction in capability.

Complement: A logical operation that inverts a signal or bit. The complement of 1 is 0. The complement of 0 is 1.

Component: *See Device.*

Computer Interface: An interface that consists of an electronic circuit designed to communicate data and instructions between a PLC and a computer. It also can be used to communicate with another computer-based intelligent device.

Conductor: A substance that easily carries an electrical current.

Conduit, Flexible Metal: A flexible raceway of circular cross section specially constructed for the purpose of pulling in or withdrawing wires or cables after the conduit and its fittings are in place.

Conduit, Flexible Nonmetallic: A flexible raceway of circular cross section specially constructed of nonmetallic material for the purpose of pulling in or withdrawing wires or cables after the conduit and fittings are in place.

Conduit, Rigid Metal: A raceway specially constructed for the purpose of pulling in or withdrawing wires or cables after the conduit and fittings are in place. It is made of metal pipes of standard weight and thickness, permitting the cutting of standard threads.

Contact: One of the conducting parts of a relay, switch, or connector that are engaged or disengaged to open or close an electrical circuit. When considering software, it is the junction point that provides a complete path when closed.

Contact Bounce: The continuing making and breaking of a contact after the initial engaging or disengaging of the contact.

Contact Symbology: A set of symbols used to express the control program using conventional relay symbols.

Contactor: A device for repeatedly establishing and interrupting an electrical power circuit.

Continuity: A complete conductive path for an electrical current from one point to another in an electrical circuit.

Continuous Rating: A rating that defines the substantially constant load that can be carried for an indefinite period of time.

Control Circuit: The circuit of a control apparatus or system that carries the electrical signals directing the performance of the controller but does not carry the main power circuit.

Control Circuit Transformer: A voltage transformer used to supply a voltage suitable for the operation of control devices.

Control Circuit Voltage: The voltage provided for the operation of shunt coil magnetic devices.

Control Compartment: A space within a base, frame, or column of the machine used for mounting the control panel.

Control Logic: The program. Control plan for a given system.

Control Panel: *See Panel.*

Control Station: A control station contains one or more push buttons, pilot lights, or selector switches in the same enclosure.

Controller, Electric: A device (or group of devices) that serves to govern, in some predetermined manner, the electric power delivered to the apparatus to which the device is connected.

CRT: Abbreviation for *cathode ray tube,* an electronic display tube similar to the familiar TV picture tube.

CRT Programmer: A programming device containing a cathode ray tube (CRT). This programming device is primarily used to create and monitor the control program. It can also be used to display data.

CRT Terminal: An enclosure that contains a cathode ray tube (CRT), a special purpose keyboard, and a microprocessor; used to program a PLC.

Current-Carrying Capacity: The maximum amount of current that a conductor can carry without heating beyond a predetermined safe limit.

Current-Limiting Fuse: A fuse that limits both the magnitude and duration of current flow under short-circuit conditions.

Cycle: An alternating current waveform that begins from a zero reference point. It reaches a maximum value and then returns to zero. It then reaches a negative maximum value and returns to the zero reference point.

Data: Information encoded in digital form. It is stored in an assigned address of data memory for later use by the processor.

Data Bus: A bus dedicated to the transmission of data.

Data Terminal: A peripheral device that can load, monitor, program, or save the contents of a PLCs memory.

Delta Connection: Connection of a three-phase system so that the individual phase elements are connected across pairs of the three-phase power leads A-B, B-C, C-A.

Derate: To reduce the current, voltage, or power rating of a device to improve its reliability or to permit operation at high ambient temperatures.

Device (Component):   An individual device used to execute a control function.

Dielectric: The insulating material between metallic elements of any electrical or electronic component.

Digital: The representation of numerical quantities by means of discrete numbers. It is possible to express in binary form all information stored, transferred, or processed by dual state conditions; for example, ON-OFF, CLOSED-OPEN.

Disconnect Switch (Motor Circuit Switch):   A switch intended for use in a motor branch circuit. It is rated in horsepower and is capable of interrupting the maximum operating overload current of a motor of the same rating at the same rated voltage.

Disconnecting Means: A device that allows the current-carrying conductors of a circuit to be disconnected from their source of supply.

Downtime: The time when a system is not available for production due to required maintenance either scheduled or unscheduled.

Drop-Out: The current, voltage, or power value that will cause energized relay contacts to return to their normal deenergized condition.

Dual-Element Fuse: Often confused with time delay. Dual element is a manufacturer's term describing fuse element construction.

Electrical Equipment: The electromagnetic, electronic, and static apparatus as well as the more common electrical devices.

Electrical Optical Isolator: A device that couples input to output using a light source and detector in the same package. It is used to provide electrical isolation between input circuitry and output circuitry.

Electrical System: An organized arrangement of all electrical and electromechanical components and devices in a way that will properly control the machine or industrial equipment.

Electromechanical: A term applied to any device in which electrical energy is used to magnetically cause mechanical movement.

Electronic Control: Electronic, static, precision, and associated electrical control equipment.

Elementary (Schematic) Diagram: A diagram in which symbols and a plan of connections are used to illustrate in simple form the control scheme.

Enclosure: A case, box, or structure surrounding the electrical equipment to protect it from contamination; the degree of tightness is usually specified (such as NEMA 12).

Encoder: An electronic, mechanical, or optical device used to monitor the circular motion of a device.

Energize: To apply electrical power.

Exposed (as applied to electrically live parts):   Capable of being inadvertently touched or approached nearer than a safe distance by a person. It is applied to parts not suitably guarded, isolated, or insulated (NFPA 70-1984 National Electric Code).

External Control Devices: Any control device mounted external to the control panel.

Fail-Safe Operation: An electrical system designed so that the failure of any component in the system will prevent unsafe operation of the controlled equipment.

Fault: An accidental condition in which a current path becomes available and that bypasses the connected load.

Feeder: The circuit conductors between the service equipment or the generator switchboard of an isolated plant and the branch circuit overcurrent device.

Ferroresonance (Transformer): Phenomenon usually characterized by overvoltages and very irregular wave shapes and associated with the excitation of one or more saturable inductors through capacitance in series with the inductor.

Filter (Electrical): A network consisting of a resistor, capacitor, and inductor used to suppress electrical noise.

Frequency: Expressed in cycles per second or hertz, it is the number of recurrences of an event within a specific time period.

Fuse: An overcurrent protective device containing a calibrated current-carrying member that melts and opens a circuit under specified overcurrent conditions.

Fuse Element: A calibrated conductor that melts when subjected to excessive current. The element is enclosed by the fuse body and may be surrounded by an arc-quenching medium such as silica sand.

Gate: A circuit having two or more input terminals and one output terminal, and in which an output is present when, and only when, the prescribed inputs are present.

Grounded: Connected to earth or to a conducting body that serves in place of the earth.

Grounded Circuit: A circuit in which one conductor or point (usually the neutral or neutral point of a transformer or generator windings) is intentionally grounded (earthed), either solidly or through a grounding device.

Grounded Conductor: A conductor that carries no current under normal conditions. It serves to connect exposed metal surfaces to an earth ground to prevent hazards in case of breakdown between current-carrying parts and exposed surfaces. If insulated, the conductor is colored green, with or without a yellow stripe.

Guarded: Covered, shielded, fenced, enclosed, or otherwise protected by suitable covers or easings, barriers, rails, screens, mats, or platforms to prevent contact or approach of persons or objects to a point of danger.

Hardware: Includes all the physical components of the programmable controller, including the peripheral.

Hardwired Logic: Logic control functions that are determined by the way devices are interconnected, as contrasted to programmable control in which logic control functions are programmable and easily changed.

Hermetic Seal: A mechanical or physical closure that is impervious to moisture or gas, including air.

Hertz (Hz): Cycles per second; the unit measuring the frequency of alternating current.

Hysteresis (Magnetic): Refers to a certain magnetic property of iron, which causes a power loss when iron is magnetized and demagnetized due to the change in magnetic flux in the iron lagging behind the change in mmf that causes it.

Impedance: Measure of opposition to flow of current, particularly alternating current. It is a vector quantity and is the vector sum of resistance and reactance. The unit of measurement is the ohm.

Inching: *See Jogging.*

Input: Information sent to the processor from connected devices through the input interface.

Input Device: Any connected equipment that will supply information to the central processing unit, such as switches, push buttons, sensors, or peripheral devices. Each type of input device has a unique interface to the processor.

Inrush Current: In a solenoid or coil, the steady-state current taken from the line with the armature blocked in the rated maximum open position.

Instruction: A command or order that will cause a PC to perform one prescribed operation.

Interconnecting Wire: A term referring to connections among subassemblies, panels, chassis, and remotely mounted devices; does not necessarily apply to internal connections of these units.

Interconnection Diagram: A diagram showing all terminal blocks in the system, with each terminal identified.

Interface: A circuit that permits communication between the central processing unit and a field input or output device.

Interlock: A device actuated by the operation of another device with which it is directly associated. The interlock governs succeeding operations of the same or allied devices and may be either electrical or mechanical.

Interrupting Capacity: The highest current at rated voltage that a device can interrupt.

I/O: Abbreviation for *input/output.*

I/O Address: A unique number assigned to each input and output. The address number is used when programming, monitoring, or modifying a specific input or output.

I/O Module: A plug-in type assembly that contains more than one input or output circuit. A module usually contains two or more identical circuits, for example, 2, 4, 8, or 16 circuits.

I/O Update: The continuous process of revising each and every bit in the I/O tables, based on the latest results from reading the inputs and processing the outputs according to the control program.

Isolated I/O: Input and output circuits that are electrically isolated from any and all other circuits of a module. They are designed to allow for connecting field devices that are powered from different sources to one module.

Isolation Transformer: A transformer used to isolate one circuit from another.

Jogging (Inching):   A quickly repeated closure of the circuit to start a motor from rest to accomplish small movements of the driven machine.

Joint: A connection between two or more conductors.

Kilohertz (kHz):   1000 Hertz.

Kilowatts (M):   1000 watts.

Ladder Diagram: An industry standard for representing relay-logic control systems.

Ladder Element: Any one of the elements that can be used in a ladder diagram, including relays, switches, timers, counters, etc.

Ladder Program: A type of control program that uses relay-equivalent contact symbols as instructions.

Language: A set of symbols and rules for representing and communicating information among people or between people and machines. The method used to instruct a programmable device to perform various operations.

Latch: A ladder program output instruction that retains its state even though the conditions that caused it to latch ON may go OFF. A latched output must be unlatched. A latched output will retain its last state (ON or OFF) if power is removed.

Latching Relay: A relay that can be mechanically latched in a given position manually, or when operated by one element, and released manually or by the operation of a second element.

LED: Abbreviation for *light-emitting diode.* A semiconductor diode, the junction of which emits light when passing a current in the forward direction.

Legend Plate: A plate that identifies the function of operating controls, indicating lights, etc.

Limit Switch: A switch operated by some part or motion of a power-driven machine or equipment to alter the associated electric circuit.

Line Printer: A hard copy device that prints one line of information at a time.

Location: In reference to memory, a storage position or register identified by a unique address.

Locked Rotor Current: The steady-state current taken from the line with the rotor locked and with rated voltage (and rated frequency in the case of alternating current motors) applied to the motor.

Logic: A process of solving complex problems through the repeated use of simple functions that can be either true or false (ON or OFF).

Logic Control Panel Layout: The physical position or arrangement of the devices on a chassis or panel.

Logic Diagram: A diagram that shows the relationship of standard logic elements in a control system; it is not necessary to show internal detail of the logic elements.

Logic Level: The voltage magnitude associated with signal pulses that represent ones and zeros in digital systems.

Machine Language: A program written in binary form.

Magnetic Device: A device operated by electromagnetic means.

Magnetostrictive Material: The phenomenon of magnetostriction relates to the stresses and changes in dimensions produced in a material by magnetization and the inverse effect of changes in the magnetic properties produced by mechanical stresses. Nickel, alloys of nickel and iron, invar, nichrome, and various other alloys of iron exhibit pronounced magnetostrictive effects.

Main Memory: The block of data storage location connected directly to the central processing unit.

Master Terminal Box: The main enclosure on the equipment containing terminal blocks for the purpose of terminating conductors from the control enclosure. Normally associated with equipment requiring a separately mounted control enclosure.

Microprocessor: A digital electronics logic package capable of performing the program control, data processing functions, and execution of the central processing unit.

Microsecond: One millionth of a second.

Millisecond: One thousandth of a second.

Modem: Acronym for *MODulator/DEmodulator*. It is a device that employs frequency-shift keying for the transmission of data over a telephone line. It converts a two-level binary signal to a two-frequency audio signal and vice versa.

Motor Circuit Switch: *See Disconnect Switch.*

Motor Junction (Conduit) Box: An enclosure on a motor for the purpose of terminating a conduit run and joining motor to power conductors.

NAND: A logical operation that yields a logic "I" output if any input is "0" and a logic "0" if all inputs are "I".

Network: An organization of devices interconnected for the purpose of intercommunication.

Noise: Random, unwanted electrical signals normally caused by radio waves or electrical or magnetic fields generated by one conductor and picked up by another.

Nominal Voltage: The utilization voltage (see the appropriate NEMA standard for device voltage ratings).

NOR: A logical operation that yields a logic "I" output if all inputs are "0" and a logic "0" output if any input is "I".

Normally Closed, Normally Open: When applied to a magnetically operating switch device such as a contactor or relay or to the contacts thereof, these terms signify the position taken when the operating magnet is deenergized. The terms apply only to nonlatching types of devices.

NOT: A logical operation that yields a logic "I" at the output if a logic "0" is entered at the input and a logic "0" at the output if any logic "I" is entered at the input.

Octal Numbering System: A number system that uses eight numeral digits (0 through 7); base 8.

Ohm (8): Unit of electrical resistance.

Operating Floor: A floor or platform used by the operator under normal operating conditions.

Operating Overload: The overcurrent to which electrical apparatus is subjected under normal operating conditions; such overloads are currents that may persist for a very short time only, usually a matter of seconds

Operator's Control Station: *See Push-Button Station.*

Optical Coupler: A device that couples signals from one circuit to another by means of electromagnetic radiation, usually visible or infrared.

Optical Isolation: Electrical separation of two circuits with the use of an optical coupler.

OR: A logical operation that yields a logic "1" output if one or any number of inputs is "1" and a logic "0" if all inputs are "0."

Oscilloscope: An instrument used to visually show voltage or current waveforms or other electrical phenomena, either repetitive or transient.

Outline Drawing: A drawing that shows approximate overall shape with no detail.

Output (PC): Information sent from the processor to a connected device through an interface.

Output Device (PC): Any connected equipment that will receive information or instructions from the CPU. This may consist of solenoids, motors, lights, etc.

Overcurrent: Current in an electrical circuit that causes excessive or dangerous temperature in the conductor or conductor insulation.

Overcurrent Protective Device: A device that operates on excessive current and causes and maintains the interruption of power in the circuit.

Overlapping Contacts: Combinations of two sets of contacts actuated by a common means; each set closes in one of two positions and is arranged so that its contacts open after the contacts of the other set have been closed (NEMA IC-1).

Overload: Operation of equipment in excess of normal full-load rating or a conductor in excess of rated ampacity, which, if it were to persist for a sufficient length of time would cause damage or overheating.

Overload Relay: A device that provides overload protection for the electrical equipment.

Panel: A subplate upon which the control devices are mounted.

Panel Layout: The physical position or arrangement of the components on a panel or chassis.

Parallel Circuit: A circuit in which two or more of the connected components or contact symbols in a ladder diagram are connected to the same set of terminals, so that current may flow through all the branches. This is in contrast to a series circuit where the parts are connected end to end so that current flow has only one path.

Pendant (Station): A push-button station suspended from overhead, connected by means of flexible cord or conduit but supported by a separate cable.

Plugging: A control function that provides braking by reversing the motor line voltage polarity or phase sequence so that the motor develops a counter torque that exerts a retarding force (NEMA IC-1).

Plug-In Device: A plug arranged so that it may be inserted in its receptacle only in a predetermined position.

Potting: A method of securing a component or group of components by encapsulation.

Power Factor: The ratio between apparent power (volt-amperes) and actual or true power (watts). The value is always unity or less than unity.

Power Supply: The unit that supplies the necessary voltage and current to the system circuitry.

Precision Device: A device that operates within prescribed limits and consistently repeats operations within those limits.

Pressure Connector: A conductor terminal applied with pressure to make the connection mechanically and electrically secure.

Program: A planned set of instructions stored in memory and executed in an orderly fashion by the central processing unit.

Programming Device: A device for inserting the control program into the memory. The programming device is also used to make changes to the stored program.

Push-Button Station: A unit assembly of one or more externally operable pushbutton switches. It sometimes includes other pilot devices, such as indicating lights and selector switches in a suitable enclosure.

Raceway: Any channel designed expressly for and used solely for the purpose of holding wires, cables, or bus bars.

RAM: *Random-access memory.* Referred to as read/write as it can be written into as well as read from.

Readily Accessible: Capable of being reached quickly for operation, renewal, or inspection without a worker having to climb over or to remove obstacles or use a portable ladder, etc.

Rejection Fuse: A current-limiting fuse with high interrupting rating and with unique dimension or mounting provisions.

Relay: A device that is operative by a variation in the conditions of one electric circuit to affect the operation of other devices in the same or another electric circuit.

ROM: *Read-only memory.* A type of memory that permanently stores information.

Rung: A ladder-program term that refers to the programmed instructions that drive one output.

Scan Time: The time required to read all the inputs, execute the control program, and update local and remote I/O.

Schematic Circuit: *See Elementary Schematic Diagram.*

SCR: *Silicon-controlled rectifier.* A semiconductor device that functions as an electrically controlled switch for DC loads.

Semiconductor: A device that can function either as a conductor or nonconductor, depending on the polarity of the applied voltage, such as a rectifier or transistor that has a variable conductance depending on the control signal applied.

Semiconductor Fuse: An extremely fast-acting fuse intended for the protection of power semiconductors.

Sequence of Operations: A detailed written description of the order in which electrical devices and other parts of the equipment should function.

Series Circuit: A circuit in which all the components or contact symbols are connected end-to-end. All must be closed to permit current flow.

Shielded Cable: A single- or multiple-conductor cable surrounded by a separate conductor (shield) in order to minimize the effects of other electrical circuits.

Short-Time Rating: A rating that defines the load that can be carried for a short, definitely specified time with the machine, apparatus, or device being at approximately room temperature at the time the load is applied.

Single Phasing: An occurrence in a three-phase system in which one phase is lost.

Software: Any written documents associated with the system hardware. This can be the stored program or instructions.

Solenoid: An electromagnet with an energized coil, approximately cylindrical in form, and an armature whose motion is reciprocating within and along the axis of the coil.

Solid State: Circuitry designed using only integrated circuits, transistors, diodes, etc.

Starter: An electric controller that accelerates a motor from rest to normal speed. (A device for starting a motor in either direction of rotation includes the additional function of reversing and should be designated a controller.)

Static Device: As associated with electronic and other control or information handling circuits, a device with switching functions that has no moving parts.

Status: The condition or state of a device; for example, ON or OFF.

Stepping Relay (Switch): A multiposition relay in which wiper contacts mate with successive sets of fixed contacts in a series of steps, moving from one step to the next in successive operations of the relay.

Subassembly: An assembly of electrical or electronic components, mounted on a panel or chassis, which forms a functional unit by itself.

Subplate: A rigid metal panel on which control devices can be mounted and wired.

Swingout Panel: A panel that is hinge mounted in such a way that the back of the panel may be made accessible from the front of the machine.

Symbol: A widely accepted sign, mark, or drawing that represents an electrical device or component thereof.

Temperature Controller: A control device responsive to temperature.

Terminal: A point of connection in an electrical circuit.

Terminal Block: An insulating base or slab equipped with one or more terminal connectors for the purpose of making electrical connections.

Termination: The load connected to the output end of a transmission line and provision for ending a transmission line and connecting to a bus bar or other terminating device.

Three Phase: Three different alternating currents or voltages, 120 degrees out of phase with each other.

Tie Point: A distribution point in circuit wiring other than a terminal connection where junction of leads are made.

Time-Delay Fuse: A fuse that will carry an overcurrent of a specified magnitude for a minimum specified time without opening. The current/time requirements are defined in the UL 198 fuse standards.

Torque: The turning effect about an axis; it is measured in foot-pounds or inch-ounces and is equal to the product of the length of the arm and the force available at the end of the arm.

Transient (Transient Phenomena): Rapidly changing action occurring in a circuit during the interval between closing of a switch and settling to steady-state condition or any other temporary actions occurring after some change in a circuit or its constants.

Triac: A semiconductor device that functions as an electrically controlled switch for AC loads.

Truth Table: A table that shows the state of a given output as a function of all possible input combinations.

TTL: Abbreviation for *transistor-transistor logic*. A semiconductor logic family in which the basic logic element is a multiple-emitter transistor. This family of devices is characterized by high speed and medium power dissipation.

Undervoltage Protection: The effect of a device operative upon the reduction or failure of voltage that causes and maintains the interruption of power to the main circuit.

Undervoltage Release: The effect of a device operative upon the reduction or failure of voltage that causes the interruption of power to the main circuit; voltage will return to the device when nominal voltage is reestablished.

Ventilated: Provided with a means to permit circulation of air sufficient to remove excess heat, fumes, or vapors (NFPA 70-1984, National Electrical Code).

Viscosity: The property of a body that, when flow occurs inside, forces a rise in such a direction as to oppose the flow.

Volatile Memory: A memory whose contents are irretrievable when operating power is lost.

Volt (V): The unit of electrical pressure or potential.

Watt (W): The unit of electrical power.

Wheatstone Bridge: A circuit employing four arms in which the resistance of one unknown arm may be determined as a function of the remaining three arms that have known values.

Wireway: A sheet-metal trough with a hinged cover for housing and protecting electrical conductors and cable, in which conductors are laid in place after the wireway has been installed as a complete system.

Wobble Stick: A rod extended from a pendant station that operates the STOP contacts; it functions when pushed in any direction.

Word: The unit number of binary digits (bits) operated upon at any time by the central processing unit when it is performing an instruction or operating on data.

Wye Connections: A connection in a three-phase system in which one side of each of the three phases is connected to a common point or ground; the other side of each of the three phases is connected to the three-phase power line.

INDEX

Page numbers in italics indicate that the entry is included in a table or figure.